On Media, On Technology, On Life

Interviews with Innovators

Editors

Arthur Clay, Timothy J. Senior

River Publishers

Published, sold and distributed by:

River Publishers

Alsbjergvej 10

9260 Gistrup

Denmark

www.riverpublishers.com

ISBN: 9788770225953 (Hardback)

9788770225946 (Ebook)

Printed on acid-free paper.

Contents

Preface

This is the second book in the series 'Interview with Innovators'. As with the first, this collection of interviews is rooted in events that took place in Seoul, South Korea as an integral part of the Seoul BioArt Festival organised by Biocon Labs of Seoul National University. The festival featured an exhibition of contemporary B ioA rt on the theme of abundance, bringing together a wide range of international artists whose work represents an important cross-section of B ioA rt output over the last decade. A parallel conference hosted talks by these invited artists on the subject of 'wet media', exploring the role of emerging technologies in arts practice, the use of living materials as an artistic medium, and how scientific knowledge (for example, of infection transmission scenarios and evolutionary processes) can ground new artistic work – together capturing a boundless vision of innovative and meaningful interactions between humans, machines, and micro organisms. A reflection on this work from Arthur Clay comes under '**Becoming Media: Yesterday's Fiction; Today's Reality**' on page 4.

Although the planning of this second volume of Interviews for Innovators originally featured only core participants from the conference, it was decided to expand its remit on the key issues being presented by inviting further artists to take part. With a larger roster of artists, it became possible to focus not only on the use of wet media in new work, but also to explore the concept of becoming media – a consequence of bringing living processes and organisms into play. With the depth and variety of work at hand, we would need a conceptual framework to mould the book into a coherent whole, whilst enabling authors to explore their own work in a way commensurate to their own vision. For this, we turned to the writing of Robert Mitchell. Mitchell points to a new understanding of living media that is a synergy between its material properties and its function as a communicative tool, driving highly dynamic and generative characteristics that can be termed *vital communication*. Through Mitchell's theory of media, it becomes possible to explore artworks not just in terms of what they are made of (or the story they have to tell) but in terms of how they reconfigure relationships between living materials, tools, techniques, people, and institutions to ask new questions of life itself. We explore this in '**Media and the Theatre of Life Innovation**' on page 8.

The book comes in interesting times. With the advent of the COVID-19 pandemic, we are faced with unprecedented challenges to our health, communities, economies, and freedom of movement. We are also seeing a new confrontation with other forms of life and the many ecosystems of which we are (only one) part. It is in such circumstances that the arts may prove an important factor in how we grapple with such issues, offering a more tactile, multi-faceted, and media-rich approach. It is through new, cross-disciplinary collaborative partnerships with artists that we can come to a deeper understanding of these challenges (what it is to truly know them) and the immediate or possible-future impacts of our interactions with other life. It lends a certain hope in knowing that human minds are a creative tool which, when put to work, can address even the greatest challenges.

Becoming Media: Yesterday's Fiction; Today's Reality

Arthur Clay

Yesterday's Fiction, Today Reality

The 21st century has ushered in an era of making the fiction of yesterday into the reality of today: contemporary artists have become designers of the future. Research in bioinformatics, for example, has attracted the interest of a new generation of artists who, through their work, propose future scenarios in which a glimpse of post-humanism is staged as an artwork and the performative role of the artist takes on a demonstrative form of participatory evolution.

To achieve such scenarios, many artists have adapted design fiction as an approach to platforming new research and use it as a catalyst to debate and explore possible futures that are foreshadowed in their work. The works are often staged as provocations, achieved by conceiving a speculative scenario on a public platform so as to share a critical perspective with the viewer. In turn, the work inspires a debate with the goal of increasing the awareness around social, cultural, and ethical issues which are truly in need of being addressed.

Of course when design fiction is used to provide a critical perspective, it is, in its essence, a form of critical design, and by adapting such a practice, artists are able to use it to directly address the survival of the human species in unique ways. Through the invention of new and inspiring imaginaries about the future, artists can aid greatly in suspending disbelief about the need for change in the present and so help shape positions that can draw the general public into participating. Through engagement with the artwork, the public is able to immerse themselves in a subject and help drive the issues at hand forward, and do so by simply reflecting on the possible consequences of current cultural values, morals, and practices.

Becoming Media

The increasing presence at exhibitions of artworks based on these principles is clear proof of the development of a new hybrid form in the arts, where artworks are staged in a theatrical manner, in which art objects are used as requisites, and where the use of living organisms and tissues, i.e., wet media, is prevalent. Taking a brief look at the use of such wet media in recent B ioA rt works, it becomes clear that a very new type of media has entered into the palette of the artist, and its inclusion brought new advances in participatory culture as well as introduced innovative ways to connect such media to existing technologies. As most of the artworks have a connection to, or are made savvy for, social media, a more diverse and larger audience emerges through a proactive use of social media platforms.

Much of what is being pointed out here can be addressed through a process that is best termed 'becoming media', which first began when media art became interactive and the viewer was placed in the role of content creator. The idea of becoming media took on new dimensions with the possibilities of deeper forms of immersion using new communication technologies and then truly culminated with the exploration of 'wet media' in the arts. The so-called *wet media* is now *new media*, and it is primarily in use amongst today's B ioA rtists who have explored it in many works as a growth media. Although the use of wet media in the arts is complex and hardly standardised, the results of its use by artists becomes apparent when one realises what is being proposed through the concept of becoming media: it goes beyond the Beuysian concept of 'everyone is an artist' to one in which 'everyone is the artwork'. If we consider the potential inherent in this (in light of the fact that wet media's main use in artwork as well as in bio labs is for culturing microorganisms), then it becomes possible to imagine that if the microorganisms were made airborne and have some form of infection property, everyone would, at least conceptually, be transformed into an artwork through mere infection. Although this might seem far-fetched, it starts taking on more validity when we consider that engineering of new organisms is a common practice in bioengineering. In fact, in a more sci-fi-like scenario, one could imagine – as Critical Art Ensemble has done – that a trans-genetic virus could impact our DNA in such a way that a transformation from human to post-human is made possible.

However, in order to go more deeply into the concept of becoming media (to explain its application in more detail), it is best to look into what is being done at present in the area of BioArt . Here, an ever-increasing number of artists are making use of wet media to create artworks around fictional scenarios, which use imaginative solutions to address issues (such as future foods, food waste, global pandemics, and so on) whilst shedding light, at the same time, on the need to address important ethical issues (arising from a world in which genetic engineering is in increasing use).

Theory to Practice

Highly relevant to this discussion is the work of one of the first to use wet media in their practice, the Critical Art Ensemble, who approaches art as a tool for directly engaging the public in a performative way with areas of science that are generally unfamiliar to the wider public. To heighten the impact of their performances, members of the Critical Art Ensemble often appear dressed as professional scientists in order to mimic workers at actual biotechnology corporations. Of course, this aids in getting the audience immersed in a fictive design world so that the artists can better address the subject at hand, and although the group's primary role is to 'edutain' the public, it does effectively show how science can be referenced in a performance and be used to create a situation in which knowledge is transferred outward to a public audience.

Several of their works have engaged areas of microbiology, reproductive technologies, genetics, and transgenics. In their work *GenTerra*, for example, the group raised issues surrounding ethics and safety in biotechnology science. Based around the game of Russian roulette, audience members were engaged with the choice of releasing microorganisms into the environment. This was done with the aid of a spinning machine resembling a large revolver, where only one of ten chambers was actually 'loaded' with bacteria. Not knowing whether the bacteria was a deadly trans-genetic virus or something harmless, the audience was forced to consider the possible consequences of releasing unknown bacteria into the environment.

In this particular scenario, it is the museum visitors who are offered the choice. However, if the chance is taken and the release button is pressed, the question comes to mind: What are the results going to be when the material is released into the open? Of course, being a fictitious scenario using only common bacteria already present in the environment, the only thing of importance to consider is the conscious act of deciding to opt for participating in the possibility of releasing something into the environment when one does not really know if the act will be a harmful one. However, if it was (although highly unlikely) a deadly virus, we could indulge ourselves with more art theory and term the wet media in use in the work as 'infectious media'. Going back to the original concept of 'everyone is an artwork', we are able to identify the performative actions in the work as the process through which the audience is transformed into the artwork itself. Regardless of whether the material was dangerous or harmful, the more important aspect of *GenTerra* is that the transformation from human to artwork takes place because the audience might well believe that they have been 'infected' with the microorganism and have become the artwork.

Taking a step backward in time, it would only be fair to acknowledge that the first generation of B ioA rtists were of paramount importance in paving the way for the creation of art that engages scientific research methods and for establishing channels of communication with the scientific community that provided the proper setting for the creation of such art. The works of the artists we feature here could hardly have come about without their cooperation with scientific research centres. It is the more recent generations of BioArt ists who have truly benefited from the practices established by this first generation and who have made use of that positioning to move the field forward by creating works in which science still plays an important role, but in which the focus is placed on adapting design fictions that guide the viewer to perceive the artwork as being embedded in a social–cultural dialogue whose content reflect the problems of today and possible solutions of tomorrow.

The 'After Information Series' by Nestor Pestana, for example, is an ongoing research project that incorporates a series of fictional narratives that the artist has translated into diverse media including illustrations, films, and performative experiments. Regardless of the form in which they emerge, all of the works are imaginaries of a post-informational era – one where biotechnologies are more common and widely accepted. For example, in one of the works in the After Information Series, the fictional community *Infumis* manipulates their skin microbiome to host bioengineered bacteria that are capable of synthesising carbon pollutants into nutrients – nutrients which are then absorbed directly into the bloodstream to nourish the body. The artist places the *Infumis* community in a real-world context and has them living beneath large traffic intersections. As a fictitious scenario (a cautionary narrative) it re-depicts non-habitable space as habitable, pushing the boundaries of the possible through the proposition that biotechnological interventions can solve real-world problems in the area of food futures and environmental pollution.

A scenario in which virology is used in evolutionary roles can be experienced in the BioArt work by Paul Gong entitled the *Human Hyena*. Through a fictional scenario, the artist proposes the use of synthetic biology to create new bacteria capable of modifying the human digestive system towards that of the hyena in order to extend the taste palette of the human to that of the omnivore s. In this manner, humans would be able to consume carrion and other forms of rotting foods so that, on the one hand, food waste can be eliminated and, on the other, food resources expanded. The proposal to accomplish this is expressed artificially in an elaborately designed fiction, one in which the artist uses a genetically modified virus to introduce genetic material into cells of the body to mutate gene production that would result in 'enhancing' the host.

This work is another example of a classic use of design fiction, in which a scenario is created so that the audience is able to decipher the meaning of the work through the actions of the performers. In *Human Hyena*, the viewer who is fresh to the performance sees three performers sitting at a table covered with diverse dishes of rotten food as contents of a meal for the performers. Each of the performers holds, and licks, an object in their hands. As the objects are neither implements for eating nor something that can be eaten, the audience is forced to ponder their use. However, it is possible to decipher their meaning through the context, i.e., title of the work, the scenario, and the actions of the performers. In the end, it becomes clear that these items are designed specifically to aid proper ingestion of the synthetic bacteria, which allow the dinner guests to consume what is being offered, i.e., food waste. All in all, the scenario instil s a sense of curiosity in the visitor who is witness to a desire in the form of a voracious appetite for the non-palatable. The ironic gestures of appetite on the faces of the performers inevitably lead us to the realisation that although the scenario is fictitious, the possibility for solutions beyond what is easily imaginable are acknowledged by the members of the audience.

Conclusion

The use of the term 'media' in the arts has changed historically, and this is due to the influence of artists who work both in the arts and the sciences or whose work stems from scientific research. In BioArt specifically, the term wet media is used and it conventionally refers to bioengineered tissues or organisms as the medium of an artwork. Through the artworks discussed above, the strategy for using wet media can be understood as an interest in exploring the real and fictive possibilities of the human body such that it is able to serve as a medium itself, around which an open discussion on a variety of topics around genetics is made possible.

Wet media is certainly new to the arts, and its use under the concept of 'becoming media' is best interpreted as a kind of embodied engagement in which the human body becomes a medium for the life-forms used in the artwork. This pushes the concept of becoming-media into a phenomena in which those engaging with such an artwork are made capable of believing their own body to have been used as media in the work and that their role as mere spectators has ceased; i.e., that they are transformed into integral part of the work and integrated into the milieu of the exhibition.

Novalis understood that all enjoyment and all taking-in is a form of assimilation, meaning that processes such as eating are nothing other than assimilation of an object into oneself. So, if our understanding of the consumption of art can be compared to a kind of eating with the senses than the notion of comprehending an artwork becomes an act of assimilation, or a kind of sublimated devouring of the external, which undergoes a transformation into something that 'becomes' part of us.

Media and the Theatre of Life Innovation

Timothy J. Senior

For anyone encountering BioArt for the first time, these are works that seem to challenge a conventional understanding of disciplinary roles (as artist or scientist), of the possible relationships between technical devices and living matter, and of the connection between concepts of life (the world of ideas) and the tangible forms that life can take. How might we grapple with these provocations? In *Bioart and the Vitality of Media*[1], Robert Mitchell suggests that we need to look beyond the readily apparent (what an artwork might be made of or the message it conveys) if we are to glimpse what really is at stake in the creation of these works. This means looking more closely at the dynamic relationships generated between living bodies, ideas, objects, and professions in the creation of new work. At the heart of Mitchell's approach is a new understanding of *media* itself: To become solely pre-occupied by whether living or non-living elements are used in an artwork is to engage with media in too limited a material sense (*media as material*); equally, to concentrate on how an artwork might 'generate debate' or 'make statements' about the world is to focus too intently on a model of cultural media (*media as communication*).

In Mitchell's view, these two perspectives only offer a partial account of artwork as media. Seen together, however, they point to a new understanding of media – one that is more than the simple addition of material properties and communicative functions. Building on the concept of *vital*

[1] Robert Mitchell, Bioart and the Vitality of Media (In Vivo: The Cultural Mediations of Biomedical Science) (Seattle: University of Washington Press, 2010)

communication, a connection between the two emerges in how artworks bring into being original and changing states within living systems. Here, living systems (whether biological or social) are understood as those always in a state of becoming something else, achieving moments of stability (what we might call identity), whilst always retaining the potential for future change and adaptation. It is this theatre of life – *of life innovation* – where the artist researcher intervenes in the creation of new work. By bringing new living beings, environments, or social states into existence, the material and communicative properties of media come to exist in a highly dynamic, generative, and changeable relationship.

Mitchell's theory of media directs us to three essential questions about artworks. The first is to ask how new work participates in the theatre of life innovation, i.e., how living processes are reshaped or directed towards new states in the work. The second is to ask how these new states come about through a reconfiguration of relationships between different elements that constitute the work, including living bodies, tools, techniques, practices, institutions, legal systems, and so on. So emerges a third question: With this reconfiguration of relationships between elements, is something newly *uncovered* that cannot be explained by prior concepts or models of understanding? Here, we might ask if new concepts are needed to make sense of these altered relationships or to ask what consequences might follow from further reconfigurations amongst elements. Taking all three questions together, the narrowness of media as material property or communicative function is replaced with concern for a work's generative and capricious nature in the theatre of life innovation.

An artwork that has living material at its core may best capture this theatre in action: Here, the driving of living material into new states creates fertile ground from which to ask these three questions of work as media. Active audience participation in the work (perhaps through influencing when and how new living states emerge) might bring us even closer to this new understanding of media such that we – as living processes ourselves – become part of the medium under scrutiny. In contrast, a painting that addresses living themes but lacks a living element *per se* (for example, through depicting genetically modified animals) might best enable the viewer to adopt a considered and critical position on the impact of biotechnology: A message is received, but, shielded from living matter itself, the theatre of life innovation remains silent.

In the years following Mitchell's work, the opportunities for artist researchers to draw on new living, technical, and methodological sources have only increased. This has generated new types of work through which we can probe Mitchell's theory of media. The boundary between artworks that engage us in, or distance us from, the theatre of life innovation is neither fixed nor clear-cut. New work that captures the advances of our time serves only to blur that distinction further. Today, for example, the *simulation of life processes* (whether through computational modelling, speculative design methods, or future-historical thinking) now compellingly explores how living processes might be transformed into new states. We can ask if, far from shielding us from 'real life' through digital or textual means (as life unfolds elsewhere), these are works that place us through their intensity and credibility firmly within the theatre of life innovation.

For this book, we have brought together thirteen artist researchers whose work animates that theatre. Through Mitchell's theory of media, we explore how individual artworks (or bodies of work) reconfigure relationships between living material, tools, techniques, and institutions to ask new questions of living processes. Recognising the vitality of media as one that demands ongoing interpretation and reflection, our discussions with artists aim to capture a current moment in their creative lives as they grapple with this space in their own terms. Whilst each artist's chapter can stand alone, there is as much that unites them together as distinguishes one from the other. As such, there are many different ways to approach the material in this book. Here, we draw out five themes that juxtapose the work of different artists to reveal the depth and breadth of the theatre of life innovation.

Microbial Assemblages

The vitality of media lies in how it brings environments or bodies into existence, uncovering new questions for living processes and the factors that shape them. Three of our artists work with microbial life to ask questions around life as encounter, exchange, duration, process, and archive: For Sonja **Bäumel**, the transplantation of microbial life from her own body becomes a means of exploring self-expansion – from a multispecies body (Me) to a flourishing of independent selves (We); for Roberta **Trentin**, nurturing microbial life in parallel to raising a family of her own gives insight into the very conditions for life itself – a means to question personal versus familial growth and the interaction between nature and nurture; for Sarah **Craske**, the microbial life of the archive reveals social and natural histories of institutions – a way of uncovering new forms of cultural and intellectual exchange.

Life Containment

Life is ubiquitous, and whilst interventions may intensify or reshape the relationships between living bodies, they may also put life itself into jeopardy – whether artist, microbe, or publics: For Roberta **Trentin**, fungal spores from home produce past its best-by date ground an artistic engagement with the microbial life that is all around us – the use of domestic wares (rather than laboratory tools) creating a new kind of artistic domesticity; for Mellissa **Fisher**, the creation of new microbial worlds sourced from her own body means the suspension of her own life (as an artist) in favour of the creative life in her own work – the use of protective casing helping assuage the public's fear of contamination from this strange new foreign body; for

Anna **Dumitriu**, Tuberculosis and the meticillin-resistant *Staphylococcus aureus* (MRSA) 'super bug' hold great promise as media but require her to undertake advanced training and to operate out of biosafety research laboratories supporting the highest levels of biocontainment in the UK.

Life Beyond Bodies

In the theatre of life innovation, it is not only living bodies that are subject to transformation into new states but cultural and social life as well. Here, we see how extensive the relationships can be between different elements in the world implicated in Mitchell's theory of media: For Vivian **Xu**, the millennia-old tradition of sericulture (silk manufacturing) is still a living one, with new connections between material, living bodies, and data elements enabling a culture of innovation; for Sonja **Bäumel**, the microbial life on our skin offers the promise of a novel interactive second membrane – a powerful non-verbal platform for revealing human encounters and enabling exchange between peoples and cultures; for Wayne **de Fremery**, oral and print traditions linked to South Korea's cultural record are fertile ground for new forms of digital object – a way of creating human-centred acts of memorialisation, political action, and public discourse.

Speculative Futures

In reconfiguring relationships between living bodies, tools, practices, institutions, and so on, new work can point to different possible future states, each with its own characteristics and consequences. Three of the artists interviewed for this book speculate on adaptive human futures in response to the challenge of food security. Proposing strategies that point to alternative and highly contrasting futures, each promises a different reconfiguration of our food industries: In Nestor **Pestana**'s work, a bio-hacker community living in isolation from the industrialised world uses biotechnology to pursue extreme self-sufficiency and self-isolation; for Paul **Gong**, modification to our digestive systems allows once-inedible foodstuffs to be consumed as part of our everyday diet – a stimulus to new forms of cuisine, culinary behaviour, and social hierarchy; in David **Lisser**'s work, the gradual development of a global lab-grown meat industry is outlined, and the possible environmental and cultural fallout of a new product – CleanMeat – imagined.

Conceptual Models

With this reconfiguration of the relationships between elements in the world by media, where does a need for new conceptual models (and further reconfigurations) arise? **Boredom research** ask whether the simulation of biological systems on aesthetic grounds can create a model for the shared understanding of the environment – and our impact on it – amongst researchers and the general public; for Nestor **Pestana**, the prospect of human modification through biotechnology takes Design beyond its current status as a discipline – a call for new models of design science fiction; for David **Lisser**, the path to a meat-free global food movement is impossible to predict – a model of future scenarios and a retrospective reconstruction of how we got there may help us imagine what is in store; for Sarah **Craske**, the impact of cross-disciplinary practice is to recast materials with a biological, social, and cultural force outside of disciplinary ownership – pointing towards a new model of transdisciplinary material.

As these works will reveal, Mitchell's theory of media helps us see how questions of life innovation are proper to us all, not just the biomedical sciences. Its ramifications for our understanding of media in its reconfiguration of people, professions, objects, and ideas make us all valid actors in the theatre of life innovation.

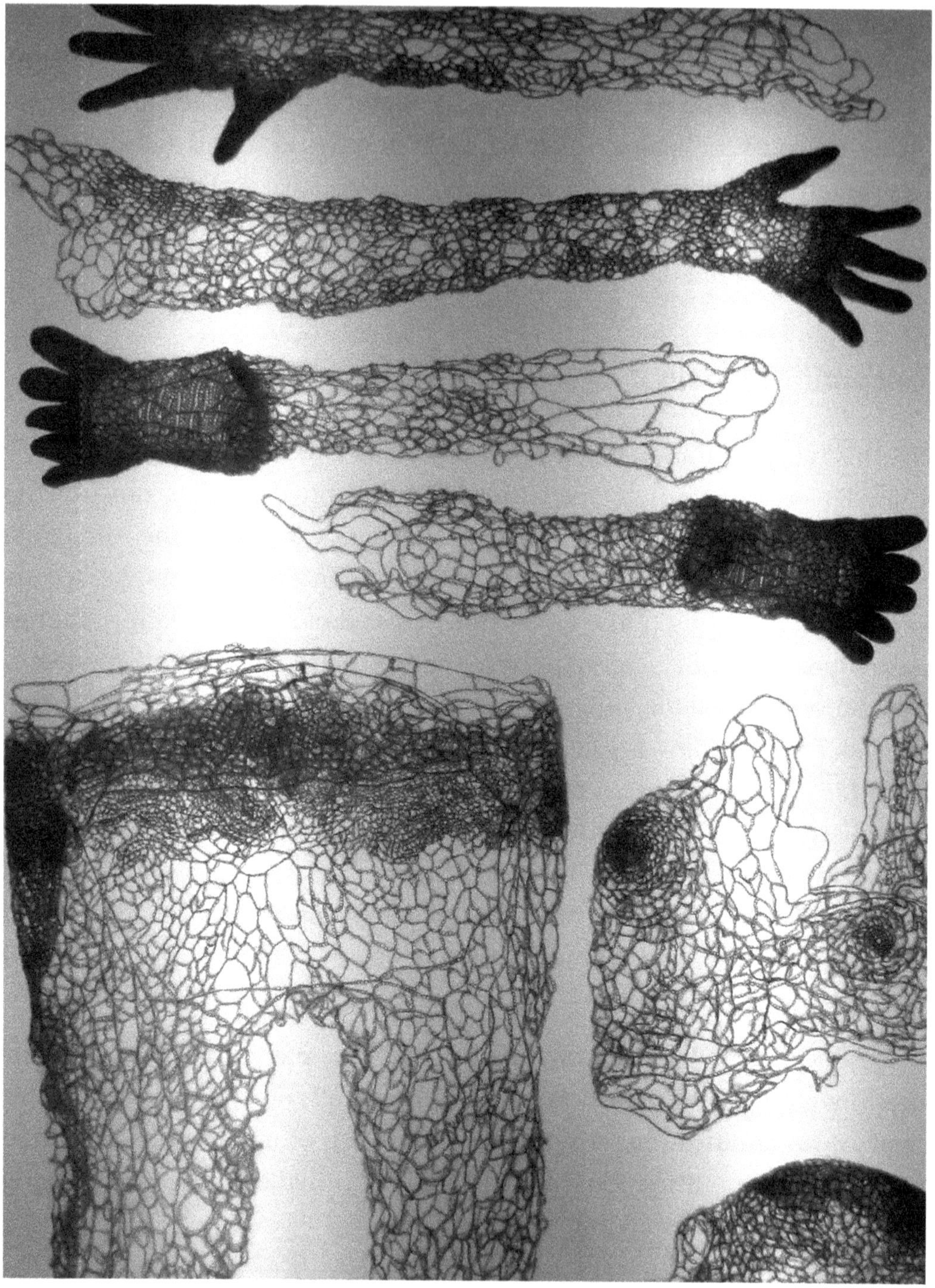

Sonja Bäumel, Crocheted Membrane (2008/2009).

Reimagining the Self

Sonja Bäumel

The human microbiome includes all microorganisms that live on and within the human body. For Sonja Bäumel, it constitutes a membrane between people – one full of life that serves as a route for exchange. In this interview, Sonja asks how this unravels our notions of the single human body, of social interactions amongst bodies, and of the legibility of human identities.

One of the startling facts positioning your work is that a large number of cells constituting our bodies are not human but bacterial, viral, and fungal. How does this force us to reimagine the human self in terms of non-human life and our wider environment?

Where does the environment begin and end? My interest lies in the microbial layer – a second skin that can be found on top of our own. It is an in-between layer, full of life, which serves as landscapes of multi-beings exchange.. The human body does not end with the skin but is continually and invisibly expanding into this fluid in-between. The in-between is full of entanglements, and our human body is just a tiny part of these microbial interactions. For more than, I have been collaborating with anthropologists, artists, cultural historians, designers, philosophers, scientists, and filmmakers to find out more about this in-between space. Moreover, I am investigating the influences scientific knowledge has had on the way we have perceived and interpreted the human body historically. My objective is transferring such an understanding within contemporary time and projecting it further into the future, striving to unravel the ways in which our comprehension of the biological body affects our current society and the cultural contexts in which we operate. I am particularly interested in how our understanding of what it means to be human is fundamentally changing in the 21st century. The democratisation of scientific knowledge and the critique of human exceptionalism are at the core of my work and its investigations into the curious relationship between humans and microbes.

Drawing on microbiologists' claims that 50% of cells that constitute our body are not human but bacterial (1), we can begin to envisage the human body as a complex ecosystem. Current research suggests that microbes might play an important role in affecting our biological and even behavioural states (such as in relation to depression (2)). In this way, the microbiome brings into focus our entangled, social, multispecies bodies. But there are also ways in which the microbiome can be used to reify old categories of distinction, and it is important to address these issues. Based on current scientific theories and discoveries revealing the influence microbes have on the human body and mind, my work seeks to stimulate our cultural imagination regarding the impact of this microbial paradigm shift: It raises critical questions about the impact this scientific research has on societal issues such as concepts of privacy, individuality, and future desires and fears; and it stages encounters with these organisms living inside and on us to explore possible futures for further co-existence. This may allow us to better take care of both the microcosm and macrocosm around us and, thus, ultimately to better take care of ourselves.

Using an example of your own work, how do these ideas play out in practice?

'Expanded Self II' is a bacterial imprint from my skin grown onto a three-dimensional agar form (the medium used in microbiology to grow bacteria) in the shape of my body. This is a work that should be understood as an expanding body in its own right rather than a singular, isolated entity. It could be seen as a metaphor offering new points of view about who and what we really are. The work was commissioned as part of the project *Gare du Nord*, which took place at the Anatomical Theatre of the Waag Society in Amsterdam (initiated and curated by Chiara Ianeselli and Lucas Evers). A quote from the humanist Caspar Barlaeus: 'Auditor, te disce; et dum per singular vadis, crede vel Iin minima parte latere Deum ' (3: p.537) ('Listener, learn yourself, and while you proceed through the individual [organs], believe that God lies hidden in even the smallest part'), written along the walls of the theatre was the project's starting point, informing the curatorial framework and interpreted by the commissioned artists, each investigating the subject of *the smallest part* from their own perspective.

I would like to emphasise the process behind this work, as it is just as important as the outcome. Visiting the anatomical theatre prompted a new development of one of my previous works – 'Expanded Self', a bacterial body imprint on two-dimensional agar. When starting a new project, I rarely have a fixed plan as I always work in a process-based fashion. I wanted to use agar, as often before in my work, so making living materials such as skin bacteria (normally invisible to the naked eye) perceivable. My aim, however, was to transform this usually two-dimensional material into a three-dimensional form, helping to better visualise the human body as a haptic and entangled landscape. This new piece challenged me, both physically and mentally, because I envisioned achieving this through producing a cast of my own body. To do this, I needed to lie for five hours in a silicone and plaster mould. The silicone mould allowed me to get all the details of the skin's surface, and the plaster mould around it helped to keep it in shape. The process revealed to me that I needed to feel my body's own borders before being able to feel its change into an expanded form.

As soon as the moulds were ready, I moved my atelier to a room next to the anatomical theatre, one I had transformed into a semi-sterile working space. Here, I worked with do-it-yourself biology (DIYbio) techniques in order to execute the work, including tools from the Open Wetlab at the Waag Society and a big pressure cooker from Mediamatic (a cultural institution in Amsterdam) to generate the large quantity of sterile agar needed. After properly setting up the space, I poured the molten agar into the mould, recruiting helpers in the middle of the night to remove the moulded body parts and place them in a giant petri dish. Once these preparations were finished, I applied the invisible bacterial collection that I had previously isolated (consisting of skin bacteria and microbes collected during a single day in a Viennese environment) onto the agar surface. The finished work was then placed in the exhibition space. Here, I started a research process – a controlled experiment – taking place within a cultural frame (i.e., an art exhibition), one where the outcome resulted from the constant transformation of this living work, a process that could not be known in advance. This gave rise to the realisation that the form we witness is not cast as a unique single piece but is the result of an accumulation of its smallest parts – microorganisms that cohabit us. Through this work, I intended to create a space where the potential of bacteria as cooperative partners can be re-imagined and where we can explore the implications for processes of larger cultural significance – art as a territory of cultural experimentation and the artist as researcher.

By placing microbial species from a shared body into a new environment (the moulded replica of your own body), you create the conditions for a further expansion of *self*. As such, how do you understand yourself to be actually embodied – but also changed – through your work?

'Oversized Petri-Dish', a work realised in 2009, was an astounding experience as it allowed me to witness a part of my own body growing independently of myself on external media. Something, which used to be part of my body, becomes autonomous and visible because the external conditions of the petri dish better favour growth (making it visible). It was a mind-blowing moment when I saw what I would usually understand as *me* suddenly flourishing as *we*. This was an important step in my practice, a kind of proof that bacteria are there and that my body is constantly expanding into space and connecting with others and the environment. The development of my body of work over the last few years has actually made this clearer to me. We are surrounded by diverse beings that are in constant exchange with their surroundings, but we cannot see or touch them. Observing the body separated from the context of its primary living organism – with most unaware of this beautiful invisibility – drives me to question our sense of body awareness. I make use of, explore, and experiment with my own body, considering it a tool to gather information about our physical and biological matter. This information is to be transformed into a comprehension of who I am and, in particular, of who we are together; this is a way of trying to better understand what bodies are really composed of. There are ethical and practical reasons for using my own body, but it also reflects my main interest in revealing the body's *bigger picture* by extending what I personally experience *through my own body* towards what happens with all bodies.

In Expand Self II, we can clearly see how you become part of the medium of your own work. In `Becoming Media', is there a risk to yourself and, perhaps, others?

Yes, I think that ' Becoming Media' in my works allows me to dive into the living material with all my senses. It is almost a proof or a necessity which allows me to tangibly feel and touch what I am thinking and talking about – a means to reflect upon ways of documenting my ongoing thoughts and to express artistic intent. I would wish to see others in our everyday lives to gain a new awareness of their bodies by focusing on their senses.

As mentioned earlier, there are very practical and ethical reasons for me using my own body in my work: First of all my body is always with me; second, there are risks involved in producing my works that only I can take. In creating 'Expanded Self', for example, I removed my natural skin flora and replaced them with a more populous, foreign bacterial layer. I had conducted a lot of experiments before producing the work in order to make sure that my natural skin flora would recover after such an intervention, though an incalculable risk still remains (as we do not know everything about microbial behaviour). I think, to be capable of fully understanding our integration with natural life-forms, we first need to gain awareness and respect for microbial life as equal partners. Our microbial populations adapt individually, their growth influenced by factors like weather, light, personal hygiene, and the use of cosmetic products. I am interested in outlining this *bigger picture* and, particularly, in understanding the ways in which we humans are connected to our surroundings and enveloped within an unknown network of other life.

It is also interesting to reflect on the issue of exhibiting living materials in the context of public spaces and traditional public institutions (e.g., museums and galleries). There is the need for a separated space (e.g., a petri dish) in order to help your living microbial material grow in the right conditions (controlling humidity, temperature, etc.) and to be safely exhibited. There is, however, still a perceived risk from audiences. My 'Expanded Self II' piece was removed from its exhibition space after a number of weeks – and prior to the exhibition end date – as it was considered by some individuals to be 'too much alive' … what a pity! Worldwide, there are very few places where you can show living artworks as they require a high level of care (these are, after all, works that are constantly growing, adapting, and changing); our conventional curation and exhibition practices are not yet ready (or elastic enough) to be able to deal with such a challenge . When you can exhibit living artefacts, it is often only within a very limited time frame, often forcing you to show only a representation of the work instead of the work itself. At the same time, when you are able to exhibit a piece, you have to begin the process of creation from scratch and reproduce the living artefact anew (which is a lot of work). This is not necessarily to your disadvantage, though, as each time you learn something more about the process, such as how new constraints or demands might influence the way the work can be exhibited.

What I have come to understand over the last years is that unless you are completely immersed in an understanding of living artworks (with its many abstract, complex, and challenging concepts), it is hard to enter into such an artwork as those I make. As these pieces have a niche focus, ask for knowledge of living systems, and largely concern media that is invisible to the naked eye, the end-result (micro-artworks) can be hard to grasp. When you want to speak to a wider audience, as I do, you need to help people directly experience the thing in question (e.g., through touch) in order to understand its own reality. Therefore, I sometimes use non-living materials in order to create imaginary landscapes that open up an understanding and conversation about life, enabling communication with a broader audience. Do not forget, we artists are still in a phase where we are trying to create an awareness and understanding for such a typology of work.

Whether in or out of a sealed environment, our everyday interactions will trigger forms of microbial exchange. Has your new understanding of the human microbiome changed the way you interact with others socially or forced you to reconsider how you see your own body in relation to others?

When I started working with microbes in 2008, I did not know about the existence of the human microbiome (such research started at around the same time). Through my work, I gained an understanding that the balance in a microbial community is very important and that the greater diversity of microbes present (of the right kind), the healthier the system is. I would say that I have become much more aware of what happens when I interact with somebody, whether hugging someone or touching the skin of other bodies, whatever those bodies are. We are part of each other. When we die, our microbes will spread to other living environments. When we breathe, we share the air and microbial life with others. We can also have a strong impact on our microbial community that is closely related to our mental and physical well-being. The microbial world inspired me as a place where we can imagine the human body as a locus of messy entangled relations, "thinking against categories such as species, sex, … , as a locus of social and biological categories in motion and in transition"[1].

So, how should we carry on our lives together? I think there needs to be a balance between maintaining the health of microbial communities and managing our day-to-day needs. For example, I would suggest trying to minimise antibiotic use where possible or reduce the use of chemical cleaning fluids which are harmful to the environment and destroy the balance of life in microbial communities. We are also what we eat, where we live, and how we care for ourselves, others, and our environment. From my point of view, the more sterile and homogenous (in terms of biodiversity) an environment becomes, the less resistance we can build. One interesting area of development that speaks to changing social behaviours is the emergence of person-to-person faecal microbiota transplant, which offers promise in restoring health to bacterial composition in the gut (4).

In how you describe your work, the idea of the *membrane* serves a very important role because of how it enables separation, transmission, and exchange. How would you describe the importance of the membrane concept in your own terms?

As you have seen, I create artefacts that blur and refashion the notion of skin. By continuing to question the way we traditionally recognise

[1] Helmreich, Stefan. 2015. Sounding the Limits of Life. Essays in the Anthropology of Biology and Beyond. Princeton: Princeton University Press. Page 62-72.

skin as a *border*, I am re-imagining skins (or membranes) as fictional layers of communication and as multi-being landscapes. After all, when we touch the surface of another organism, our skin picks up many new microbes whilst leaving others behind. In relation to the research I am conducting into individual perceptions of the human body, the fact we each have a constantly changing microbial layer surrounding our bodies could lead to the imagining of different types of communities that we share in common. In the long term, this could strongly impact not only the way humans interact but also the way societal systems and networks of other life forms are understood to operate. If only we were capable of seeing these relationships not as dividing and fixed but as more fluid, connected, and context-driven, I believe that this would enable us to better integrate with other life-forms. This may well engender more sensitive, or attentive, ways of interacting with each other and with our surrounding environment.

I have a background in fashion design, and these ideas originally emerged out of a critique of the fashion system. I started with a simple question: Do we expect too little from our clothing? It felt very natural to me to express my ideas through the use of textile. In the project 'Crocheted Membrane', the clothing I developed (or, better, body forms) do not derive from shapes or historically patterned forms with an embedded social hierarchy and pre-established material richness but are instead determined by the needs and sensations of an individual human body, performing much in the same way as bacterial populations individually respond in their environments. Nowadays, I am more interested in working with these ideas in detachment from the fashion field, so exploring the in-between space that surrounds and connects all multi-beings bodies and my interest lies in non-verbal forms of communication.

In 2009, to come back to your question,, I talked about the second secretive layer of life on our skin, which is, in fact, very close to the function of textiles and clothing. 'Crocheted Membrane' has offered me a way to materialise an understanding of the abstract and invisible biological layer that surrounds our bodies and to create an imaginary entry point for the general public to encounter such complex and challenging notions. Furthermore, through developing such work, I wanted to express a new visual language, one that could connect different audiences, such as fashion designers, scientists, and the public.

You have suggested that the ability to intervene in the microbiome might one day enable designers to tailor individual human skin bacteria populations into visible, flexibly adapting membranes. These membranes could change the function we assign to clothing, perhaps as indicators of status, health etc. How might such a membrane change the way we interact socially, and to what ends?

The '(in)visible' film project was artistically rooted in the world of fashion where critical and project-oriented approaches are rare. Although fashion is a social construct perceived in the form of clothing, the world of fashion and the world's actual needs seem strangely disconnected. Two problems I see are that, first , our clothing does not support the social and individual human body but rather celebrates conformity; second , whilst the use of forecasting does exist to generate new fashions, it is driven by prediction (i.e., based on past fashions) and commercial needs rather than the power of individuals' imaginations. I wanted to bridge this gap and redefine fashion's function as that of providing added value to the world as an accessible, fictional tool for critical reflection. In essence, my aim was to unsettle the *present* rather than predict the *future* and to critique consumer culture where profitability is its main goal. Here, design can act as a medium for the public to ask questions about the future rather than simply provide solutions to current problems.

In my first film, '(in)visible' (2008/2009), I envisioned how a novel second skin layer involving organisms such as skin bacteria, slime mould, and plants might mutate when adapting to different environments and contexts. I was interested in how we could use this powerful non-verbal platform between humans in a more flexible and meaningful way, i.e., how a novel layer on our skin could help create new forms of communication between humans within their environments. Through such work, I asked questions such as: What might clothing or a second skin look like when developing out of someone's socio-physical needs and individual beliefs? How might interactions between humans change according to the adaptive behaviour of this novel layer on our skin? Could we become more sensitive or attentive to such interactions if they were to prompt changes in colour, shape, and structure of this flexible, visible membrane? Could social integration, for example, be supported by this kind of aesthetic and functional adaptation to the environment? In imagining what would happen if our second skin was considered as something other than the way it is today, we can ask how it could sharpen our sense of our surroundings, reminding us that we are just a tiny part of a wealth of microbial interactions. A(n) (in)visible membrane could tell a story about us as human beings, about our present state, our strength, our fragility, our fears, and our feelings. New societal landscapes could evolve related to the visualisation of these characteristics.

Developing the capability to visualise the hidden interconnections between humans and microbes is the first fundamental step, acting to stimulate public awareness of these possibilities. But it is also essential for us to gain a deeper understanding of how such interactions might function in order to learn how to use them in the best way. Trying to interpret the soundless language spoken by millions of barely perceivable entities living on our skin is an essential step to building a new, living, and adaptable system on our bodies. To be capable of fully integrating with natural life systems, we first need to recognise the many ways in which microbes act as equal partners in living systems. Such a cultural shift could allow us to embrace their 'expertise' derived from billions of years of co-existence with us. Microorganisms, for example, have adapted incredibly successfully to radical changes and transformations in their environments, reflecting their capacity to evolve in relation to changing needs and pressures. We could probably learn a lot from our tiny co-habitants. The overall intent of this project was to think of broader and long-term solutions in a sustainable society, which respects human resources and environments and takes responsibility for our actions. In my more recent work (such as 'Expanded Self II'), I am beginning to explore these issues through my interest in the relation between bodies in multi-being landscapes.

As a designer of objects that embody abstract and challenging concepts, your artistic practice throws a new light on concepts derived largely from scientific inquiry. What do you understand this relationship to be, and is there an influence you would seek on the scientific community?

I often work in scientific labs together with scientists when developing my projects. I feel that by working together, by asking different questions of each other, and by observing each other's working methods, we already mutually influence each other's ways of thinking. Hence, artistic practice makes a contribution through affecting the approach and focus of the scientific community too. An art project can offer the scientist freedom – a means to exchange knowledge and gain a different perspective on knowing, understanding, and questioning the world. Although I love to work with scientists, I have never been interested in becoming a scientist. If you can gain knowledge from both fields, you can see and appreciate the world around you even more. I have now worked for many years with scientists. At first, I was always the one to initiate projects and to reach out to the scientific community for collaboration. Now, after almost ten years, I have been asked for the first time to be part of a scientific research project, contributing to the work from an artistic standpoint. It has taken time for art to gain (once again) both trust and respect from science; and, despite incremental positive steps, we are still not fully there.

It is important to ask where work on the microbiome challenges conventional understanding of the body, and where it reinforces existing distinctions and forms of discrimination. In each project I develop, I have had the opportunity to ask questions along these lines and, perhaps, make a contribution to ways of thinking in the scientific community. For instance, in my project 'Fifty Percent Human', we sought to create greater awareness around the ethical issues of how to conduct experiments with microbes whilst still finding out more about them: How we relate to other living organisms – whether a cell, a plant, or an animal – is a question of care, which means both for them and ourselves. For me, microorganisms are not simply inert DNA; I am more interested in their personalities and their behaviour in context. This means that rather than looking at a single species in isolation, a new emphasis emerges on how microbes communicate in their social context. I think I have also had a more practical influence on the scientific community. In 'Expanded Self', I transformed everyday scientific materials into novel three-dimensional shapes and scaled them up considerably, broadening the view of what can be done with these methods and possibly inspiring or informing new working practices. I believe that artists and scientists share a lot in common (for instance, the joy and intuition of experimentation), though we must be aware of key system differences. For instance, scientific work is bound to a particular way of conducting research and publishing findings, whilst, as an independent artist, I am able to have a completely different view on things: With my work, I want to critically question a variety of subjects, to provoke discussion, and to express or celebrate a different, but equally important, kind of language – that of the magical and non-verbal language of material, colour, texture, and form.

Beyond the scientific community, your DIYbio workshops enable a participating public to encounter living works in public spaces. What is the role of these workshops, and how do you think it affects participants' understanding of bacteria as operating between artistic medium, dangerous contagion, and everyday fact of life (visible or otherwise)?

The aim of these workshops was to show that science is not necessarily as complex as it seems (depending on which level you operate at, of course), to spread DIYbio know-how, and to democratise scientific knowledge in action. A workshop allows you to bring important topics into the public sphere, to provoke questions, and to engage in discussion with a diverse professional community. Furthermore, workshops act as a platform to explore how we all need to take responsibility in dealing with topics which affect our shared future, whether we want to or not. I am, for instance, referring to the issues around antibiotic resistance or the field of synthetic biology. I think that we should all have access to basic knowledge in regard to such topics, to allow all of us, ideally, to collectively take part in discussing the role of such developments and what kind of future we would wish for: This needs to be a dialogue between artists, scientists, curators, health politicians, policy makers, people from the pharmaceutical industries, and different publics.

References

[1] Sender R, Fuchs S, and Milo R. Are We Really Vastly Outnumbered? Revisiting the Ratio of Bacterial to Host Cells in Humans. *Cell*. 2016;164: 337–340.

[2] Johnson K V-A, Foster KR. Why does the microbiome affect behaviour? Nat. Rev. Microbiol. 2018;16: 647–655.

[3] Barlaei C. Antverpiani Poematum pars II, Elegiarum et Miscellaneorum Carminum. Amsterdam: Joannem Blaeu; 1645.

[4] Hota SS, Potanen, SM. Is a Single Fecal Microbiota Transplant a Promising Treatment for Recurrent Clostridium difficile Infection? *Open Forum Infect Dis*. 2018;5(3). Available from: doi:10.1093/ofid/ofy045.

Author Biography

Sonja Bäumel studied Fashion Design at the *Fashion Institute of Vienna*, and holds a Bachelor in Arts from the *University of Arts of Linz*, as well as a Master in Conceptual Design in Context from the *Design Academy Eindhoven*. Her work has been exhibited internationally in *Ars Electronica Center*, *Anthology Film Archives New York*; *MAK Museum of Applied Arts Vienna*; *Museum of Contemporary Art Taipei*, *Museum of Natural History Vienna*, *Seoul Museum of Art* or or *Centre Pompidou Paris (FR)*. She is co-founder of the Dunbar's Number collective (2011), member of Pavillion35 (2012) collective based in Vienna, and of the WNDRLUST (2013-2018) collective based in Amsterdam. Sonja is currently heading the *Jewellery-Linking Bodies* Department at the *Gerrit Rietveld Academie* in Amsterdam, as well as lecturing and giving workshops in different national and international academies and universities.

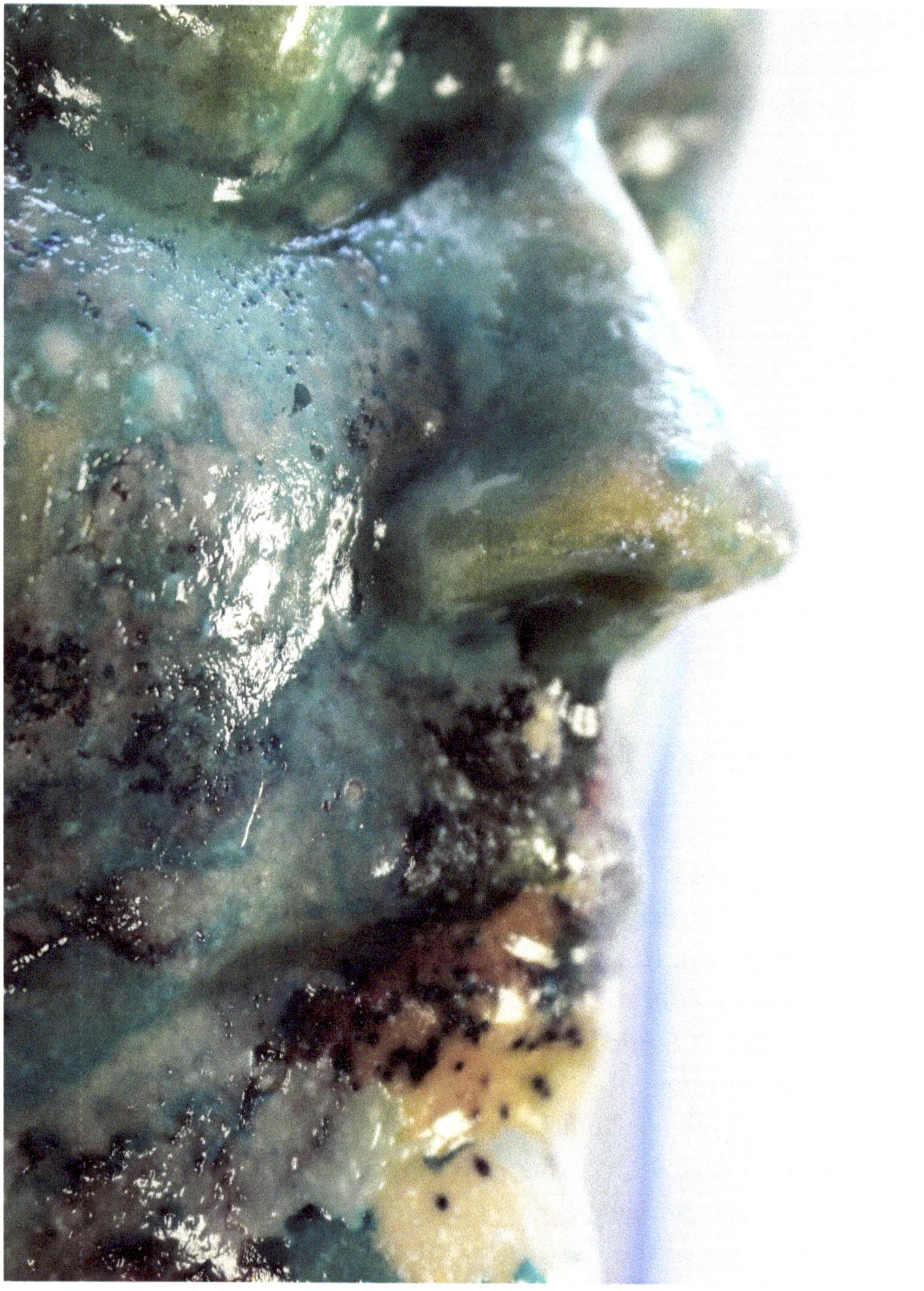

Mellissa Fisher, Microbial Me (2013).

The Invisible World and the Visible Self

Mellissa Fisher

The human body is a landscape on which hosts of microorganisms co-exist, grow, interact, and compete. This microbial body is largely invisible – a hidden part of our unique fingerprint. In this interview, Mellissa Fischer asks what happens when that body is liberated for all to see; a form of self-emancipation, for sure, but also a new beginning for her microbial self.

A key moment in your development as an artist has been the discovery that rich forms of life normally invisible to us can be made visible through scientific methods. Could you describe something of this process of discovery?

It really began during an Art and Science Interdisciplinary module I was taking back in 2010 called Broad Vision, run at the University of Westminster by artist Heather Barnett. At that time, I was an illustration student who was struggling to find inspiration in my subject, and, in short, when I looked down the microscope during that course, I found it: The shapes and colours I saw simply blew me away, and the thought of this beauty being invisible to the naked human eye convinced me it was this *invisible* world that I wanted to bring to the public through my project work as an artist. Throughout the Broad Vision course, I experimented with different ways of making the invisible world visible and found that the most fascinating approach was to render the invisible *physical*. Dr. Mark Clements, who I met during Broad

Vision, became my collaborator, and I remember asking him whether I could make sculptures out of agar. He responded: 'I have no idea', but he was interested in trying. So we tested the idea out with various moulds and realised that it worked; the structure of the agar held perfectly and retained a high level of detail. I then began to explore casting parts of my body and produced a face for my first exhibited 'Microbial Me' work (originally titled 'Face of Truth') at GV Art (London) in 2013.

'Microbial Me' is a project about the use of scientific materials in an artistic context as well as about the exploration of the microbial life found on the surface of the skin. By re-presenting skin-sourced microbes on an agar sculpture taking the form of my facial profile, I am recasting the self-portrait as a living microbial portrait – one that evolves over time. In the work, I am assigning agar, as a medium, a wider purpose that extends beyond the 2D petri dish into a 3D landscaped form. 'Microfloral Femunculus' was an extension of 'Microbial Me' – a miniature of the human body cast in agar that would bring this work closer to my original artistic intentions. This was an experimental piece in visualising the body; we wanted to test how microbes swabbed from each area of the human body might behave on a smaller, corresponding agar structure of the human form. In order to explore these behaviours further, we tested three different types of agar support medium separately, generating results that would inform later work with the medium. Our plan with this work was to start small with the body figure and work towards a method for casting a full-scale human figure.

Our everyday lives take place in near-ignorance of our own microbiomes. What do we gain from making this invisible part of our lives visible in this way? How should it alter our sense of what constitutes *self*? Indeed, has it altered your sense of self?

Mark and I drew inspiration from the initial reaction of the public when seeing 'Microbial Me'. We heard many people saying: 'I didn't know that we had bacteria on our faces' ; we were shocked by how little the public knew about these bacteria that accompany us through life and play an important role in our everyday health. Brought to our attention, this marked the start of a longer journey for us – one exploring how to render the invisible world visible for more people through bringing artistic and scientific practices into partnership. Working to alter people's *sense of self* in this way has been a phenomenal and fascinating experience for me. My interest is in a form of science communication that can educate the wider public about the human microbiome: How many bacteria, and which types, live on our skin? What is the extent of their growth over our bodies and in the environment? Just how much do we need them in order to stay alive?

In general, I think people are scared of the unknown, and so I wanted to bring the unknown to the surface to start a discussion on what it means to be human, i.e., to ask whether the self we present to others (and perceive ourselves) is really the *whole* self we are. Since I have begun working with my own bacteria, my sense of self has changed greatly. The mere understanding that bacteria are growing all over my body has shifted my self-perception towards that of a living composition made up of millions of tiny organisms; it has forced me to question what being a human really means and what kind of organism my body actually is. I even behave differently towards myself now because of this understanding. For example, I no longer obsessively clean my hands or body as much as I used to; my knowledge of bacteria has made me much more conscious of their vital role in my continued health and existence. In short, my new understanding about bacteria has made me think differently about my own mortality and my relationship with nature: I am nature, and we are nature.

In this vein, the recognisable component of the artist – the agar form moulded directly from your own face – becomes less visible (even distorted) as new microbial colonies grow. Is this the emancipation of your own microbial self, or does the eventual decline of this new ecosystem reveal a deeper set of dependencies that sustain our integrated relationship with nature?

The sculptures of my face only bear a passing likeness. Over time, I see them turning into something completely different again. As soon as they become covered with bacteria, it is no longer my face that I recognise at all. One piece (developed with Mark and Dr. Richard Harvey) has been exhibited at The Eden Project in Cornwall for almost three years; it still looks as beautiful and interesting as it did after three days. It is, technically speaking, my face, but it is a different version of my face created by my own bacteria in their own time. To the public, it is a generic face, but one similar to their own, and, therefore, one they can relate to. It could be seen as powerful in this way: The face is what people first turn to in an encounter – the first thing people look at in each other – for reassurance, for approval, and to detect emotional states. Of course, the work also resembles a death mask, introducing a tension between the suspension of my own life (as the artist) and the beginning of a new microbial world.

So, yes, the work is an emancipation of myself, but also a new beginning for my microbial self. 'Microbial Me' generates an ecosystem all of it s own based on the microbes from my skin, continuing to change and morph over time in unexpected ways. It is no less subject to processes of living and dying. In their natural environment, bacteria on our skin are in constant competition with each other (a process also replicated on the agar sculptures). The colonies can compete with each other for nutrients, with those bacteria able to grow at low nutrient concentrations becoming more dominant as the sculpture matures. Bacteria also compete with each other in more aggressive ways, such as producing antibiotics which can kill other types of bacteria or alter the environment, for example, by producing high concentrations of acid which can prevent other bacteria from growing . On our bodies, the bacteria are finely balanced; each body part will have slightly different micro-environments which favour one species more than another. This competition between bacterial colonies on the sculpture is similar to the complex interaction of human societies: Different societies compete with each other for resources (such as food and water) in the same way as bacteria do. There is also symbolism in the bacterial production of antibiotics – the equivalent to human warfare. What was a sample of bacteria living in a balanced ecosystem on my skin becomes a new ecosystem outside of my body with its own, unpredictable fate.

From another angle, the vibrant forms of microbial growth that emerge in your work are fascinating and repellent in equal measure – kept at a distance from us by a protective casing. How have audiences reacted to this tension between insight into our natural histories and the perceived risk of contamination in your work?

When the public views the work, I have noticed that there is a strong response of disgust. That seems to be the general feeling people have towards bacteria: An indifference to whether their impact on us is good, bad, or unknown. Usually, I display the sculptures in glass or Perspex casing, which allows the viewer to see the sculpture whilst sheltering them from the horrendous odour that the bacteria generate, and shielding the external environment from the risk of possible contamination. Since it is not known exactly which types of bacteria have been harvested from my skin for growth in the sculpture, all bacterial sources are treated as potentially 'dangerous'. Our future plans include sequencing the bacteria so that we can determine any contamination risk from the outset. We have faced many challenges in exhibiting the pieces: They need to be safely displayed within airtight casings and with minimal risk of being disturbed or knocked over. Although we explain to curators the best way to display the work, many have been reluctant to include them in exhibitions (seeming not to have properly understood the risk assessments that we have already undertaken).

These pieces are always seen as 'grotesque' to begin with by audiences because they are something unfamiliar, and we have all been raised to believe in the value of cleanliness – a sterile world without bacteria. What I am trying to do is help audiences see that bacteria are, naturally, everywhere and that their role in sustaining life is much more complex. Recently, I was able to work with the BBC presenter Michael Mosley on a new bacterial sculpture cast from his own body (a project I will discuss later), spending time with him over the course of its development. The sculpture made him uncomfortable; it was simultaneously exciting and disgusting, especially as it was 'himself' he was seeing down there covered in bacteria. But, over the duration of filming, he became more amazed by his own microbiome and how his bacteria were evolving to resist the broad spectrum antibiotic we applied to part of the sculpture. This initial sense of disgust is not something I worry about; I am still exploring new ways of exhibiting parts of the microbiome that can help draw-in and educate audiences.

Although an interaction with these microorganisms in your work is prevented, you raise the point that we exchange microbial life through our everyday interactions all the time. Your work makes something of this process visible, but are there other ways in which this everyday exchange outside of the laboratory (or gallery) might be made shown, and to what effect?

I explore some of these interactions through the workshops I run. In 'Design Your Own Microbiome', I ask participants to draw a self-portrait and use a marbling technique over it to create microbial patterns of the kind you would find under a microscope. Another way Mark and I plan to reveal something of our everyday microbial exchange outside of the laboratory is to sequence the microbiomes of participants, revealing their microbial fingerprints for comparison. Activities such as these continue to be important to me because they allow an engagement with the public through my practice that is safe and avoids any of the risks associated with exposure to living bacteria. None the less, I am currently developing a 'Design My Microbiome' workshop with a collective called 'BIO.CHROME' – one where participants are given casts of parts of my body onto which they apply their own bacteria, so raising questions about bacterial ownership and origins: Do all the bacteria on our skin belong to us? Do we share bacterial species, in which case which ones? How much variation can be found within our microbiomes?

Turning to questions of practice, you have engaged with research scientists and arts organisations in the creation of your work. If working with living materials offers new opportunities to explore questions around living processes, do you understand your work as exploring a topic that necessarily defeats disciplinary boundaries?

I first experimented with agar at home, although without nutrients, to work out what sculptural qualities it might have to offer. Working with agar containing nutrients essential for supporting growth, however, can only be undertaken in a lab setting; this is due to the potential risk of growing pathogenic bacteria. When I began working with agar, I was concerned principally with questions of appearance. As each type of agar used in scientific research contains indicators to reveal certain types of bacteria, I was inspired to mix two or three different agar types together to see if this would affect bacterial growth and variation – this was certainly evident in 'Microbial Me'. Mixing agars to get the desired colours, textures, and growth, I was little concerned with the application of scientific method.

My own practice of blending together different agars for artistic reasons has led me to really interesting outcomes and deeper insight into how different organisms in the microbiome respond to their environment and interact with each other. Mark and I coined the term 'bastardising agar' when experimenting in the lab with this technique (because we are not using the agar as intended but, rather, to create a new purpose for it). Some might criticise this approach, but I do not believe that cross-disciplinary working means that, when an artist and scientist collaborate, they have to focus on questions of a scientific nature. Many artists have this focus; I am more interested in experimenting with concepts and materials and newlines of artistic questioning.

Recently, I was able to expand 'Microfloral Femunculus' for the BBC Four documentary 'Michael Mosley vs. The Superbugs'. Mark and I were commissioned to create a life-size bacteria sculpture of the presenter ('Microbial Michael') to be part of this documentary on antimicrobial resistance. To make this possible, we created a new method of body-casting, one where an immovable cast of the sculpture is placed in its final orientation and then filled up on the inside with agar (from the bottom to the top) to form the sculpture. The challenges we faced with this project mainly concerned building a casing for the sculpture which could meet strict health and safety requirements whilst being aesthetically pleasing. This was a very experimental piece, and, unfortunately, the seal within the casing failed after four days, dramatically shortening the length of the time-lapse film we could create to document the work. In this process, Professor Sheena Cruickshank (my collaborator) observed how fascinated she had become with the piece: It s rich visual appearance has now inspired her to ask more questions about the types of bacteria the sculpture supports and their extended life course. In contrast, some scientists I have worked around in the lab have criticised my playful approach for not being scientifically rigorous. But I argue that I am not trying to conduct scientific research: I am trying to make the invisible world visible by experimenting with materials and pushing the boundaries of casting and sculptural form.

Let us think about the relationship between our complex real-world and laboratory practice for a moment: In these sculptures, different species of bacteria or fungi will become dominant overtime and will continue to grow until they run out of a specific nutrient or produce toxic b y-products that eventually prevent them from growing (or even kill them). This prompts a new wave of growth from another bacterial species favoured by these conditions. This process will repeat itself over and over again until all the nutrients are completely used up (which will take a very long time). Scientists are unable to predict exactly how this will occur or when the end will finally come. This is simply because they would no t normally leave an experiment for this length of time, and normally they work with pure cultures of bacteria (rather than complex communities such as bacteria from the skin). This is something where, perhaps, only working with artists such as myself will help us uncover answers to these questions – although the challenges of running a 20 year artistic experiment would be considerable!

In addition to your work with living materials, you are an active illustrator, conduct microbiology research, and have an interest in stop-frame/time-lapse animation. Is there an interaction between your work with living forms from the microbial world and these other aspects of your work?

Since working with organisms through collaboration with scientists, my artistic work has changed substantially in all areas. My interest in the representation of the invisible and the patterns it creates is now present in my illustration work (as I recreate the microbial sculptures in my line drawings). Although the microbial world has come to influence all aspects of my current work, the theme of nature and the living has always been key to my practice in some way. When starting out as an illustrator, I always wanted to communicate movement through inanimate objects; this is present in my early work with fractal patterns which served as a kind of optical illusion of the fractal equation. My engagement with the microbial world has pushed this interest in capturing *time* in my work much further. For example, my interest in seeing how bacteria grow (at different rates and in different patterns) has resulted in a number of time-lapse films. So, I think that my practice has not necessarily changed at heart, but it has evolved, like a cell dividing and reproducing – and it will continue to do that.

Another project of yours, `Immortal Ground', sees your work expanding to encompass other notions of ecology and life. Could you tell us more about this piece and describe some of the challenges you face in taking your practice forward? How might your conception of living materials continue to change?

'Immortal Ground' was a project for my final degree show – 'Unfolding Realities' – in 2016 at Central Saint Martins in London. This project originated with a residency run by artist Alexis Williams in Ottawa Canada, under the title 'Art Ayatana – Biophilia'. This residency explored themes in biology and art through various activities like hiking in Gatineau park to forage for mushrooms or learning about caterpillar interactions and cell communication. The act of foraging and being connected to nature in a way that I had not engaged with before inspired me to create the project Immortal Ground. The mushroom that I became particularly interested in was the Reishi mushroom, which in Asian culture is known as the 'immortal mushroom' because of its role in increasing the macrophages in your white blood cells and boosting the immune system. The work gave me the opportunity to explore different ecosystems and engage with medicinal plants; it brought me to think about my sense-of-self with nature and the immortal values we ascribe to the natural world.

Thinking to the future, funding is one of the biggest challenges I have to overcome in creating microbial projects. They simply cost much more than traditional projects of a similar scale as the scientific equipment required to produce the work and the protective housing needed to surround the sculptures are so costly. We now know, however, that this work is possible – we have made it happen. Finding the right environment to keep and exhibit such works is also something we are trying to resolve (itself a subject for future funding). As the body sculptures have an estimated lifespan of at least twenty years, we would like to recreate a project like 'Microbial Michael' and take the piece to 'full-term'. Working at this scale has greatly altered my perception of working with living materials. It is ambitious to create living sculptures at such a scale, especially when gallerists are anxious about exhibiting such pieces and scientists fear being part of this kind of collaborative project. I have been very lucky with my current scientific collaborators as they understand what I am trying to do as an artist, so they want to be a part of my exploratory project work; after all, it helps them to think differently about their own research.

Artist Biography

Mellissa Fisher's practice brings together interests in illustration, printmaking, sculpture, and living organisms to make the invisible world around us more visible. She holds a degree in Illustration and Visual Communication from The University of Westminster, UK. In 2016, she graduated from Central Saint Martins in London with an MA degree in Art and Science (a course that investigates the contemporary and historical contexts of artistic and scientific practice). Since 2016, Mellissa has undertaken major commissions for The Eden Project in Cornwall, UK ('The Invisible You: The Human Microbiome' 2015 –2020) and the BBC documentary 'Michael Mosely versu s the Superbugs' (first shown on BBC4 in May 2017). Mellissa continues to collaborate closely with leading research scientists in her work, and she regularly delivers participatory workshops and public talks exploring the world of art and science. More on her work can be found at https://www.mellissafisher.com/

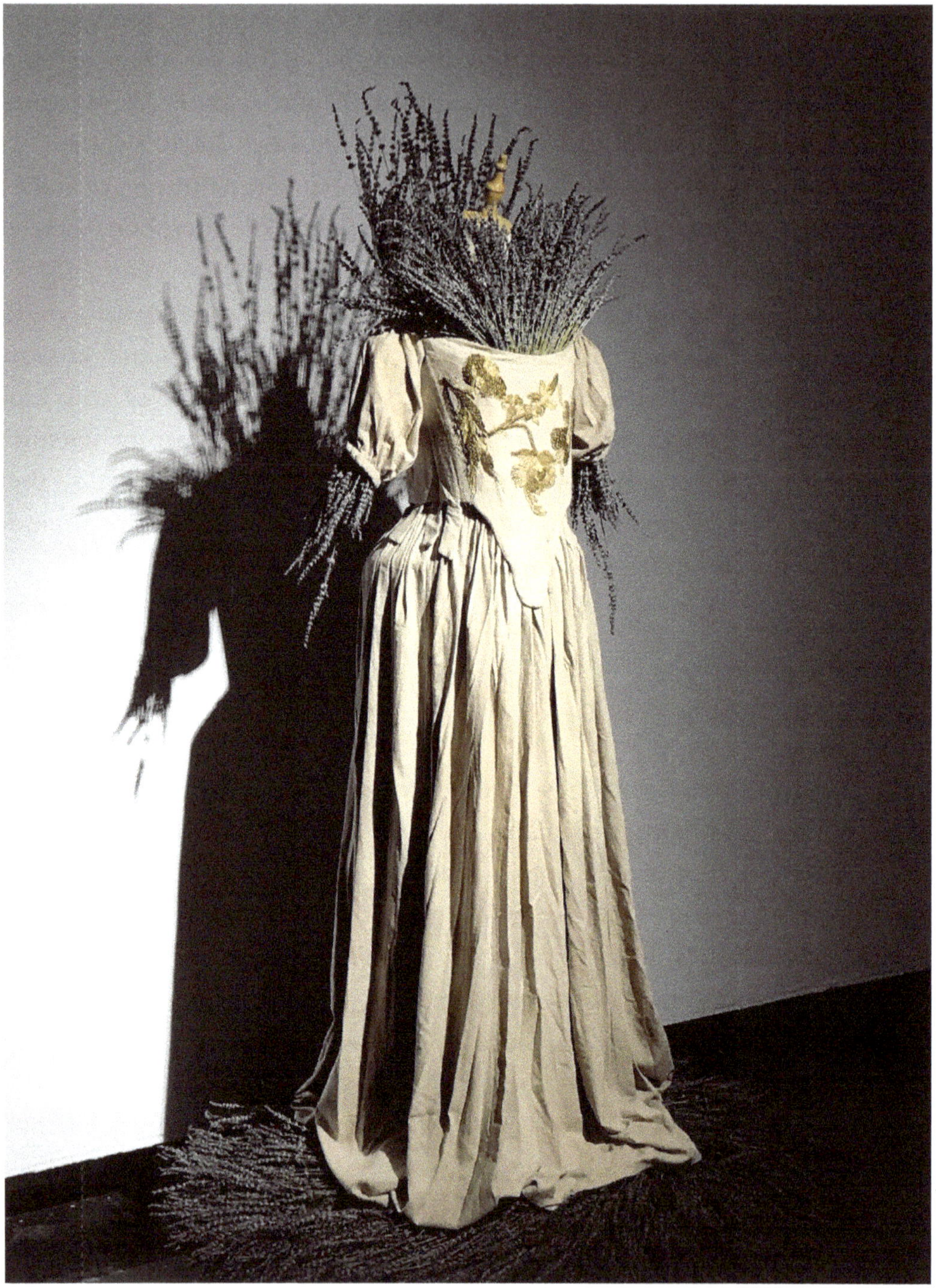

Anna Dumitriu, Plague Dress (2018).

Hidden Stories of Awe and Terror

Anna Dumitriu

For Anna Dumitriu, there is both awe and terror in the impact of bacterial life on human health. This bacterial sublime is one that always invites, but ultimately resists, our full comprehension of it. In this interview, we explore what it means to work at the forefront of collaborative practices that might just have serious consequences for your health.

Working with living media is an essential part of your work. That relationship, however, is far from utilitarian: You have invoked the idea of the `bacterial sublime' as an expression of that engagement. What is it about working with living materials that drives you?

I work mainly with bacteria. In fact, they are central to my interests in human health and disease. Bacteria are wonderful, complex organisms, and the more I learn (indeed, the more that Science learns) about them, the more fascinating they become. The notion of the bacterial sublime combines the feelings of terror and awe that we feel when reflecting on the impact these minute organisms have on human life (an impact we are only now beginning to understand). It draws on a tradition of valuing terror as an aesthetic pleasure in art and nature originating with Edmund Burke's classic text 'A Philosophical Enquiry into the Origins of Ideas of the Sublime and Beautiful' (1). The fact is that the full impact of bacterial disease on humanity is only now emerging, as whole genome sequencing allows us to look at the minute changes disease interaction has made to our genomes since the dawn of mankind. Our changing behaviours – such as our original descent from the trees, our domestication of animals, our changing diet (from herbivore to omnivore), and forced or voluntary large-scale migrations – have driven an exposure to unfamiliar diseases or zoonoses (those that can be naturally transmitted between animals and humans), transporting them into fresh populations of 'victims' that lack any form of immunity . But it is wrong to think of disease as something outside of ourselves – as the 'other'. In fact, our co-evolution with disease is an integral part of what it means to be human.

Through an intense focus of working artistically with bacteria, I have built up a level of experience and expertise that enables me to work with infectious organisms. This is important as these are the bacteria that most significantly affect humanity – there are so many stories to be told about them. Using the same artistic methodology, I explore themes from microbiology, genomics, and synthetic biology: They all are part of a spectrum of research focussed on understanding the nature of life. In this way, I am interested in revealing hidden stories and investigating our impact on the natural world; I am interested in drawing out threads across time – from our history to our potential futures. Personally, I have worked with many different kinds of bacteria – from extremophiles that live in extreme environments (such as highly polluted sites or the Arctic Tundra) to dangerous organisms including *Mycobacterium tuberculosis* and MRSA . When I work with dangerous organisms, I collaborate with suitable laboratories to ensure all necessary safety and biocontainment requirements are met.

Through my work, there have been many important discoveries and learning experiences – both for me and the scientists around me. A real breakthrough in my own work came when I was collaborating with Dr. James Price: We discovered that we could impregnate textiles with bacteria and use chromogenic-selective agars to support their growth, and so pigment the cloth; by using things like antibiotic discs and silk embroideries treated with natural antimicrobial dyes, we could then alter these bacterial growth patterns. We found that these textiles retained their colours when sterilised, which meant I could create artworks for public display that reveal how various treatments for infectious diseases work and show how chromogenic dyes can be used to diagnose different forms of bacterial disease. Another discovery was made during the development of my 'Engineered Antibody' synthetic biology work when my collaborator Xiang Li (University of California Irvine) commented: 'Working with Anna on the antibody necklace piece actually made me realise that I had an error in the sequence of my antibody that I am using in my research project. [To build the work] we had to compare my antibody sequence to the correct antibody sequence in a crystal structure, and I noticed that those sequences did not match. Since then, I have fixed the sequence of my antibody for my research project !'. This shows how working with an artist can force a kind of 'quality check' in science because I make my collaborators explain everything they do until it is clear to me – a necessary step if I am to make a practical, physical artistic response. By working with me to make an artwork, Xiang realised something did not sit right, and that is how he discovered his error. It takes an artist who does not just sit back when they do not understand something for such a situation to arise. I have built up my knowledge in the field over many years, and so I am able to engage with it quite deeply. I think it is more than just creating a space for reflection though – it is about working practically on an artwork that makes you think in different ways.

Many artistic practices involve living material that can be readily generated in a home environment. Your practice, in contrast, has also brought you into direct engagement with MRSA and Tuberculosis DNA, amongst others. In what is essentially a field of innovation, how has your work demanded new types of collaboration between different disciplines, institutional activities, legal frameworks, and so on?

I work closely with trained microbiologists and scientists, i.e., embedded in laboratory settings, and have done so for many years (working in this way since the late 1990s). I collaborate in this way not only because it is necessary but also because these interactions inspire me. All sorts of things drive me in this work, not only the physicality of the media we work with (the bacteria, the agar, etc .) but also the conversations we have whilst working in the lab, whether about the history of bacteria, experiences of conducting experiments, or new research that is planned. All my work is made by me (hands-on in the lab), so I make sure I am compliant with all legal requirements (and undertake all necessary health and safety training) during the research and development stage of my work; I cannot develop and exhibit new work unless it is safe to do so.

I have learned 'on the job' so to speak ; so my knowledge of this field has developed over many years, with lots of support coming from the scientific community. In this way, I have learned about the interrelated legal, health, and safety aspects of this work, and how, for example, regulations differ between countries. There is legislation governing biocontainment in the UK, for example, that means you need a separate license to work with genetically modified organisms; this makes it very difficult to exhibit live, genetically modified bacteria outside of the laboratory. My knowledge and experience of working with these organisms has also evolved as research has evolved. In fact, I have experienced the development of sequencing technology in infectious diseases first-hand through my collaboration with the 'Modernising Medical Microbiology' project, seeing directly how it has impacted our understanding of the mechanisms of infection and epidemiology, and how it has led to an explosion in the field of synthetic biology.

Regulations can impinge on artistic intentions. Sometimes the display of such organisms requires a certain form of containment or regulation, and in these cases, I work with all necessary parties to understand what is needed and ensure that it happens. These include the scientific collaborators, the senior supervisor of the research in question, the curators involved, the venue, and even, potentially, representatives from Environmental Health or the Health and Safety Executive. All have been very supportive in my experience, working with me to ensure that I can carry out my work whilst ensuring all regulations and requirements are maintained.I rather enjoy the challenge of trying to get work shown, and, in those cases where I try to do things that have never been done before, I engage the opinions of lots of scientists over how to do it safely. Recently I have been trying to develop sculptures containing wild antibiotic resistance plasmids – mobile elements of DNA that contain genes that provide bacteria with a kind of upgrade to be resistant to antibiotics. I say 'wild' because they are from the environment rather than a lab-engineered plasmid. Normally, the lab-engineered plasmids do not confer resistance to the antibiotics used in human healthcare. Legally speaking, displaying DNA is not a problem, but it is not clear if there is a risk that bacteria in the environment could take up these resistance genes and become superbugs. Since I first proposed the artwork, the wild antibiotic resistance plasmids that I want to use have appeared in the UK population and so are no longer something we risk releasing. Without bacterial hosts, plasmids are hard to put into bacteria outside the lab, so transfer into a suitable bacterium (if present) would be extremely rare, if it occurs at all. In fact, the answer is not yet known to Science.

Your work is conducted in a safe environment, with measures put in place to minimise risk to both yourself and others. Nonetheless, what is the experience of working with pathogenic microorganisms?

It is as simple as working in the correct types of labs, with their normal health and safety requirements, and with the correct types of bio containment. Occasionally, I have been offered, or required to have, vaccinations: For example, to work with faecal samples from patients, I had to be vaccinated against Typhoid. There are three categories of labs for the handling of bacteria (numbered from one to three), with level-3 supporting work with the most dangerous kinds of bacteria. I have worked with bacteria up to biocontainment level 3 and have received plenty of training: This has included lab inductions (some very intense), training courses, hands-on experience, and a lot of 'on the job' instruction (2). I have, for example, worked with *Yersinia pestis* (the organism that causes plague), which requires a very secure category 3 lab; I have not yet made any actual artworks with this medium but aim to do some work with it at some point soon – the development of my work sometimes takes many years.

Nowadays, modern lab procedures dominate the experience of working with bacteria. For me, *Yersinia pestis* is one of the most sublime bacteria, but when I was able to work with plague the first time, the processes associated with lab work overtook any aesthetic sensation of the sublime I was seeking. Instead, a whole range of other sensations overwhelmed me – from a sense of being privileged to have entered this space and to share it with others, of clumsiness (or fear of clumsiness at least), to inadequacy, but also a sense of achievement. Some sense of the 'bacterial sublime' is still with me, though, every time I step inside a microbiology lab, and it is an experience that I need to share through my art practice.

In all instances, any pathogenic quality in your work has been extinguished prior to an encounter with the public. How do people respond to your work? Does a lingering doubt as to their own safety remain – a memory of the living so to speak?

I do not think it is so much a lingering doubt about safety, as it is understood that all potential pathogens have been killed. Where traces of these organisms remain (e.g., in how they have grown on cloth and stained it), they taint the object with that history – sometimes invisible and sometimes quite clearly showing their growth and interactions. The traces of these organisms play out artistically and relate to philosophical notions of the sublime, such as Kant's position that an experience of the sublime is situated in the mind of the observer rather than in the object, although the object is the trigger: There is nothing intrinsically sublime about cloth and bacteria alone, but together there arise interactions that can stimulate extraordinary imaginative possibilities.

Of course, it is not possible to say how much of the intention of the artist is included in the experience of the sublime; although I try to make my work affecting to the imagination, it depends on the viewer and how the work triggers their experience. To some extent, the works can also be appreciated on a 'retinal level' as aesthetic objects (I want my works to be visually impactful, after all), but I think the viewer cannot help but read some of the references I layer into my artworks. My artworks all have a strong conceptual sub layer to them that informs their initial creation and shapes their aesthetic impact. Although I cannot speak for all artists working in this field, I think it is quite common to focus on such conceptual elements. This is because we, as BioArtists, often work with invisible things that need to be made visible in some way. Ethical issues – often hidden in the work of the Sciences – are, for this reason, frequently explored in BioArt.

If an engagement with such work is heightened when the material used is still living (and, therefore `generative' in some way), how far would you like to go in creating an artwork that brings an active pathogen and the general public into direct contact?

There is a connection between biosafety levels and the concept of the bacterial sublime: An element of terror is very important for a sublime experience, although such an experience depends on the viewer's sensitivities. I have always wanted to create a biosafety level-2 lab in an art gallery. This would enable visitors (after the appropriate training and assessment) to participate in the handling of living pathogens and genetically modified organisms. An art gallery setting of this kind would enable visitors to engage in this experience aesthetically, attuning them to the sublime rather than allowing the structures of science to wholly supervene on their experience (as a scientific laboratory setting might). In such a setting, a visitor would only differ from a participant in whether they enter the lab or just observe it from the outside. The training needed would address more than just legal and health and safety issues: I think help in tuning into the aesthetic aspects of the experience would be an important part of the overall training process. The question of whether Science supervenes on the experience of viewing BioArt is something I explore in 'Confronting the Bacterial Sublime' (3) – part of my long running art/science ethics project 'Trust Me, I'm an Artist' (4).

What mainly stands in the way of creating a biosafety level-2 lab in an art gallery is that it is very expensive and that I have not found a suitable funder yet. Saying that, I have now worked with scientists to enable the display of (killed) pathogens in a science museum setting. Let us say I am still working on the plan, although its final instantiation might well evolve as my ideas evolve. Several people have suggested that using biosafety level-1 organisms in a gallery-based biosafety level-2 lab might be something that would be of benefit to both me and my work, whilst also being of educational value to a participating public. However, there would be little point to doing this, as the cost would still be very high for such a simple artifice. After all, a biosafety level-2 lab would allow me to work not only with pathogens but with genetically modified organisms as well; to these ends, my current efforts are to establish a 'gallery-lab' that would allow me to work with both.

Reflecting on this further, the artistic practices you describe are often used to challenge our understanding of the ethics surrounding artistic and scientific work. What are some of the lessons you have learned in this exploration of ethics?

All BioArt is very much tied up with ethical issues. Often an exploration of the ethical implications of a certain technology or practice becomes the subject matter of the work or at least informs the subject. The project 'Trust Me, I'm an Artist' aims to teach artists how to deal with ethics committees so that they can make and display work without causing harm to themselves or the public: It helps curators understand how to support this process and exhibit work; gives them the tools to understand the implications of such work; enables institutions to feel more confident in exhibiting BioArt; and gives ethics committees advice on how to successfully work with artists. As part of 'Trust Me, I'm an Artist', we ran a series of performative ethics committees, using an event structure that I developed with Professor Bobbie Farsides to reveal their real inner workings. It is not so much a case of asking whether lessons have, or have not, been learned but about developing a shared journey towards a consensus on best practice; this is something that is continually developing. Often the things we explore, for example, my interest in displaying wild antibiotic resistance plasmids in my artwork, have never been tried before, so raise questions for which there are no clear scientific answers at present. At the moment, I am trying to find out how to make this particular project a reality, but as the idea behind it is somewhat ahead of scientific knowledge, I may well have to do the scientific experiments myself. Sometimes I think what we need are new ways of taking such questions forward, as they can fall into a crack between established forms of artistic and scientific practice. In the end, though, these sorts of collaborations are really mutually beneficial, with artists often raising new research questions, helping researchers to reflect on what they are doing, bringing ethical issues and debates to the foreground, and suggesting or proposing new uses for emerging technologies.

What is your favourite ethics committee biography from this project?

'Trust Me, I'm an Artist' was triggered in part by my own experiences with ethics committees, but also by a conversation I had with Neal White about his 'Self Experimenter' project – a subversive re-enacting of Yves Klein's 'La Vide'. The original piece consisted of an emptied exhibition space guarded by sentries painted blue; its contents were obscured until the space was entered. Non-invited participants were charged a large amount to enter the space; here, they were served Methylene Blue cocktails. Neal's piece focussed on these cocktails, which apparently would cause participants to have blue pee the next morning (a private artwork for them to enjoy). Concretely, Neal wanted to offer people Methylene Blue pills that they could take at their own risk – in light of research showing Methylene Blue can cross the blood –brain barrier. He wanted to perform the work in a medical research facility but was advised that it was not ethically possible to do it there. Instead, a member of their ethics committee recommended that he perform it in a gallery, where it would be permitted. There is a connection between self-experimentation or self-exploration, in the arts, but such work is no longer supposed to take place in the sciences.

With much of Klein's work, he tried to make his audience experience a state where an idea could simultaneously be *felt* as well as *understood*. The development of such conceptual strategies is important in bringing ethical issues to light for the public. I think I respond most to BioArt works that stem from this Fine Arts perspective; it is a kind of work that has a way of drawing the public into wider debates and different ways of thinking. We actually managed to do a 'Trust Me, I'm an Artist' event with Neal White, which I have written about and documented in my book. In the same vain, I really enjoyed events with Adam Zaretsky and, more recently, with Kira O' Reilly and Jennifer Willet, events that questioned the relationships we draw between species and environments – the laboratory as a natural ecology and the wilderness as a laboratory. They each raised interesting ethical concerns, explored issues of biocontainment in different ways, and questioned current research practices in Science.

In the work you describe, there is often a foreshadowing of new relationships between disciplines and practices demanded by changing times. What are some of the most interesting, far-horizon challenges that may come to preoccupy artists, designers, and scientists in the future?

A lot of my work is about drawing threads across time from the history of science and medicine to emerging technologies and paradigms. What seems clear to me is that our understanding changes very rapidly and that medical procedures, scientific beliefs, and ethical approaches are often in a huge state of flux: What seems logical at one time can seem ridiculous or barbaric even twenty years later. One far-horizon challenge that is particularly interesting to me, and highlights this point, is the future of antibiotics. A number of my pieces look at issues around the current and future impact of antibiotic resistance, exploring the consequences of how we have misused antibiotic drugs since their discovery. This is particularly relevant to tuberculosis care – artificial pneumothorax (a treatment to collapse the lung) used to provide a 'rest' cure for tuberculosis patients prior to the advent of anti-tuberculosis medication; strangely, with the present issue around antibiotic resistance, it may be that we will need to look again at such treatments. Two works of mine that look at this issue are 'Make Do and Mend' and 'The Hypersymbiont Dress'.

'Make Do and Mend' (5) references the 75th anniversary of the first use of penicillin in a human patient in 1941. It takes the form of an altered vintage wartime woman's suit marked with the British Board of Trade's utility logo CC41 ('Controlled Commodity 1941', meaning that the use of materials had been deemed to meet the government's austerity regulations). I patched the holes and stains in the suit with silk patterned with genetically modified *E. coli* bacteria. These were created with Dr. Sarah Goldberg using a cutting-edge technique called CRISPR (Clustered Regularly Interspaced Short Palindromic Repeats) allowing researchers to cut and paste DNA. By removing the gene responsible for resistance to the antibiotic ampicillin and then scarlessly patching the bacterium's DNA to encode the World War II slogan 'Make Do and Mend', we were able to 'mend' the organism back to its pre-1941, pre-antibiotic era state. The suit is accompanied by a series of framed works combining WWII CC41 textiles, altered WWII leaflets that inspired the piece, relics from the CRISPR experiment, and a child's toy sewing machine (that my mother used to play with during WWII) shown stitching a silk grown with modified bacteria. In making that piece, it seemed somehow right to include this toy, although that decision was made quite instinctively. The theorist Annick Bureaud has raised the point that the use of the sewing machine suggests we are still 'playing' with these techniques and that we are still not sure what the consequences of their use might be.

The second piece – 'Hypersymbiont Dress' (6) – plays with this idea further by asking how new technologies might enable forms of interaction with our own bacterial flora (or even foreign infectious diseases), a move that could enhance us as organisms and drive our evolution at a cellular level. This project has involved extensive collaborative work, first with Kevin Cole and Dr . John Paul, Dr. James Price, and Dr. Rosie Sedgwick, then further work with Alex May, Dr. Daire Cantillon and Professor Martyn Llewellyn. The piece takes the form of a dress both stained and video mapped with forms of bacterial life that could turn us into human super-organisms – with improved creativity, improved health, and even improved personalities. The dress is stained with normal environmental bacteria, but also *Mycobacterium vaccae* (a soil bacterium that enhances cognitive function by increasing serotonin levels, as tested and proven in rats), MRSA (which can interface with the human nervous system and affect how we feel pain), and *Bacillus Calmette Guerin* (BCG; a form of attenuated Bovine

Tuberculosis that has been strongly linked to creativity throughout history). The video mapping on the dress was created from a film of my own blood fighting, *in vitro*, an infection with BCG.

It is undeniable that the development of new technologies carries potential risks and may lead to unpredictable consequences. However, I do not think art has to be about identifying or solving those problems; for me, it is about raising deeper questions about what it means to be stupid, fleshy, rotting bodies facing the world that confronts us, enabling us to reflect on the complexity of our biology, its aesthetics, and our failure to fully comprehend it. In my work, I want to give people tools to think critically about what they read and hear in terms of new technologies, to be able to tell the hype from the reality, and to provide a way of understanding scientific and technological ambiguities rather than just expecting black and white answers – the world has few of those.

References

(1) Dumitriu A. *The Bacterial Sublime*. Available from: https://annadumitriu.co.uk/portfolio/the-bacterial-sublime/ [Accessed 25th September 2020].

(2) Dumitriu, A. Trust Me, I'm an Artist: Building Opportunities for Art & Science Collaboration through an Understanding of Ethics. *Leonardo*. 2018;51(1): 83–84. Available from: doi:10.1162/LEON_a_01481.

(3) Dumitriu A. Trust Me, I'm an Artist #3: Confronting the Bacterial Sublime: Building a BSL 2 Lab in a Gallery. Available from: https://vimeo.com/45938252 [Accessed 25th September 2020].

(4) Dumitriu A. *Trust Me, I'm an Artist*. Available from: https://annadumitriu.co.uk/portfolio/trust-me-im-an-artist/ [Accessed 25th September 2020].

(5) Dumitriu A. *Make Do And Mend.* Available from: https://annadumitriu.co.uk/portfolio/make-do-and-mend/ [Accessed 25th September 2020].
(6) Dumitriu A. *The Hypersymbiont Dress.* Available from: https://annadumitriu.co.uk/portfolio/the-hypersymbiont-dress/ [Accessed 25th September 2020].

Artist Biography

Anna Dumitriu is an artist working at the forefront of collaborative practice between the arts and sciences, focussing primarily on human health and disease. Her installations, interventions, and performances use a range of biological, digital, and traditional media; these include live bacteria, robotics, and textiles. Her work has been shown at The Picasso Museum in Barcelona, ZKM Center for Art and Media in Karlsruhe, The Science Gallery Dublin, The Museum of Contemporary Art Taipei, The Guangdong Museum of Art, and The Victoria & Albert Museum in London; her work is held in several major collections, including the Science Museum in London and the Eden Project, UK. She is Artist- in-Residence on the Modernising Medical Microbiology Project at The University of Oxford and with the National Collection of Type Cultures at Public Health England. More on her work can be found at http://annadumitriu.tumblr.com/

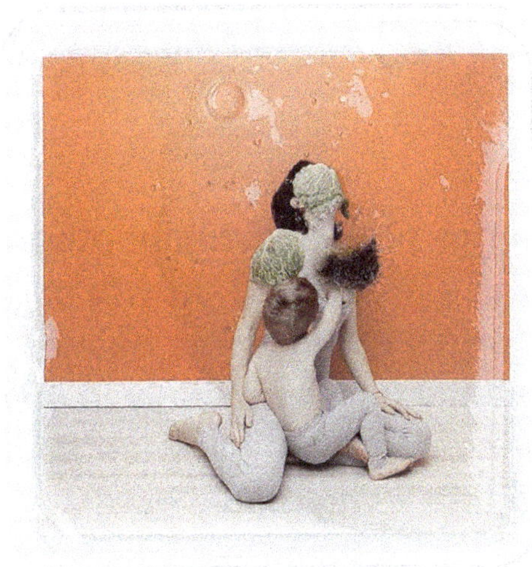

Roberta Trentin, Bios/βίοσ series (2012 –2015).

A Narrative of Warmth

Roberta Trentin

Life is ubiquitous, but none is closer to us than our own. Sometimes a consideration of life in general, or in the form of `the other', can uncover what might normally be taken for granted. In this interview, Roberta Trentin asks how a juxtaposition between microbial and familial growth creates a fulcrum around which the passage of time, values, and nurture can be observed.

In your Bios/βίοσ series (2012–2015), photographic portraits capture your family in a process of growth and transformation. The parallel that emerges in the works between microbial and familial life is an unusual one, especially in such an intimate context. What is the comparison you wish to draw between the notion of family and this living media?

When a family is created, a new entity is born, and different layers are built on top of each other. These layers consist of the emotions, hopes, needs, and efforts that propagate within that new entity. Atop these elements, there is the pivot of all – time. With time, the interactions within the new entity become more dynamic, and time translates into co-existence: there is harmony, collaboration, understanding, learning, but also conflict and tension. With time, the entity grows because the individual components do. Further, just like any living organism you might experience growing, a family needs warmth, nurturing attention, space (nurturing my own as a parent and allowing it to my kids to be what they like to be), and protection. Lastly, there is a macro concept here around survival, a central driving motivation for all species: Mould reproduces through spreading spores just as we perpetuate our species through the creation of the next generation, each subject to evolutionary pressures and the action of natural selection. These are the foundations that help us thrive as a family and unfold as values that will propagate with the new generations within the family. Growth is brought on by a passage of love and deeply held values.

The juxtaposition I use between microbial and familial growth to explore these ideas is now a critical one in my work. You would need to look at my work from the year 2012 up to the present to really grasp the concept of growth that I am trying to invoke; the *Bios/ βίος* series is revealing in this way. Under development since 2012, the project looks at the concept of growth within a family – *my own family*. During my year at the International Center of Photography in New York, I came across the work of Elinor Carucci; I loved her aesthetics and the way I could feel the warmth of motherhood and family. Inspired by her approach, I started to explore the narrative of my own family and the deeper connection between myself, my partner, and the child we 'created'. For the first time, I had a clear feeling of wholeness built by unique and single parts. That is when I thought of us as a 'microorganism' colony, an entity for the first time! In *Bios/βίος*, having a living organism superimposed onto family portraits is the key to showing the different concepts of growth I am interested in. When I apply mould to the plate, which contains nutritive agar, it takes only up to a few hours for the mould to start growing. The medium and the warm temperature of my home generate good growth conditions for mould. The plate itself becomes a home for growth, in parallel to my own family home. Lastly, a still photograph captures an exact moment in time, whereas the presence of the growing organism on top of the still photograph renders the work a truly dynamic and lively ensemble, suggesting familial growth in the background. Physical, conceptual, and emotional layers come together; their two-dimensional stillness loosens up and allows flow to take place.

The human microbiome is a unique identifier for us individually and collectively. It is, in part, acquired through birth and subject to our daily interactions with others. Bios/βίος 2014 captures you at a late stage of pregnancy; it seems to speak both to this microbial connection between mother and child, but also of danger, contagion, and infection?

While I was pregnant, I found myself thinking profoundly about my growing womb. There was growth within growth, and, cell after cell, a new life formed. That was the reason why I placed the mould on top of my belly in that work. Also, I wanted to reveal and showcase my particular beliefs about mould. People are used to looking at mould as something contagious and unhealthy, and often associate it only with rotten and decaying food. In some cases, it is true that it is unhealthy, for example, when you are co-existing with black mould. However, my understanding of mould is that it is an organism that lives and grows; its life cycle interacts with the surrounding environment and – like all living matter – it manifests under certain conditions. Mould grows spore after spore. Once a spore lands on a surface, it searches for water and nutrients to feed off. As the spore takes root, it begins to spread and create more spores and spread quickly over the surface. This process easily reminded me of the pregnancy stage of conception and implantation of the human embryo into the womb.

Another thing that I have often heard regarding my work is the term decay. However, the only decay that I perceive from my work is the one represented by my growing family. We start ageing the moment we are conceived and the life cycle applies to us just as simply as with any other living organism (and no matter how complex the organism is). Imagine *Bios/βίος* in 10 years, and you will probably be able to grasp what I am trying to explain! I believe that decay is part of the growth that we face, and although it is commonly associated with death, I strongly see it as the opposite. In nature, decomposition and decay are vital processes as they allow organic matter to be broken down, recycled, and made available again for new organisms to utilise. In my own family, as we grow, we make available experiences that will *mould* our children into persons that will create their own families one day. Isn't death the completion of a life lived?

Unlike some artists who work with biological media in laboratory conditions, you source your fungal spores from home produce and use kitchen oven gloves to prepare agar plates in the home. Does your work speak in this way to a kind of domesticity, an everyday existence with microbial life that largely goes unnoticed?

My work as an artist is mainly inspired by what happens in the sanctuary of my own home. Domestic realities are my biggest inspiration, so you are absolutely correct. My work has a kind of domesticity about it and is really about everyday life and interactions with each of us as individuals and a family unit. At the moment, I am actually working on three other projects that take their inspiration from the kitchen counter, daily routines with my children, or simply being home in the most secure place. However, generally speaking, my family's co-existence with microbial life happens both inside and outside of the home.

In *Bios/ βίος*, I have really come to embrace the fact that I am not working under the controlling and sterile conditions of a bio lab; in contrast to a lab setting, I expect much more than just what I put onto my petri plates to come out and show its presence. In my laboratory work, I was trying to prove the validity of a technique that sterilises fruit fly males such that their release into the natural environment will reduce the fitness of the overall population. My work focused on researching the impact of the sterilisation method on the bacteria present in the fruit flies' head and gut. In order to get reproducible results, my working environment had to be as uncontaminated and sterile as possible.

However, the domestic environment in which I create my art is nothing but uncontrolled – no matter if I follow the same procedures very strictly, the end results are often unpredictable, creating irreproducible pieces of art. I look at the presence of bacteria in my plates like the things we cannot control in life, things that happen outside our immediate family but that still relate to us and have an impact on our emotional balance and health – whether in the working environment, at school, amongst extended family and friends, but also in the political and spiritual sphere. These external factors can be either positive or more challenging, but, none the less, they exist and they influence us as a family. This way of creating work frees me to explore growth in a way that reflects what happens in real life. So, in direct opposition to the thinking behind my lab work, I embrace and incorporate unplanned occurrences into my art, speaking up for that unpredictability that life reserves for us – the unpredictability that is a key part of life itself.

The parallels you draw between the growth of living entities, between different forms of life, and between practices of art and everyday life extends the medium of the work beyond the mere confines of the sealed agar plate. Is there a sense, then, that in engaging with these works, we too `become media'?

There is, indeed. The parallels that are drawn in my work between the growth of living entities and the processes that connect them are deep within my practice. Certainly, the connections extend the medium beyond the confines of the agar plate and/or the paper printed photograph. Concerning the idea that we, too, become media, I think that the act of using a camera (re) creates us as media in the work, especially given that there is, from my side, an intentional performance for the camera as well. I do not direct my kids or my husband with a great amount of details: I let them know what I think the pose should portray and I let them express themselves freely. With these portraits, the layers of mould, agar, and the petri dish itself (all built on top of each other) aim to create a sense of living media as well. Furthermore, in *Bios*, the translucent features that agar possesses not only transforms the concept of canvas but also offers a new ephemeral layer for the artist to engage with. Agar becomes a surface of convergence between the concepts performed and those shown. The viewer is called to get closer to the piece, to look at the details, grasp the presence of life, and perhaps to experience growth with the same deep sense I put into creating the work.

Could you tell us more about the photographic element of the work and the interaction between photo and living media over it – from the emergence of air bubbles in the agar to patterns of emphasis and concealment caused by microbial growth?

I welcome the unpredictable behaviour of the living media as it emphasises the fact that in life, we too face a good deal of encounter with the unknown. From our life experiences to relationships, we can actively influence our life's own path, but we often have to leave room for unforeseeable events. However, the air bubbles are an active element of the medium itself and I am really glad you mentioned them. First of all, the bubbles add a wonderful texture to the two-dimensional image and give an extra spark of life as well. Second of all, agar plates that contain bubbles are not very helpful when studying bacterial colony growth; bubbles make the counting and identification of bacteria colonies much harder, so it is very important not to have too many of them. Bubbles also add a playful touch that, when raising young children, becomes a daily routine in the family. Working in a lab where I had to prepare hundreds of plates in a day, the most stressful part was to make sure there were no bubbles; and I mean not even one. However, in my art practice, I can finally free myself from that and allow unpredictability to be a part of my art.

Bios/βίοσ has unfolded over four years. As the project and your own family have grown, have the parallels around the concept of *growth* generated new insights, connections, or associations beyond the original expectation of the work?

Bios/βίος is an ongoing project – a constant observation of how we grow as individuals and as a family. The work has witnessed an important conceptual shift over the years, allowing room to reflect on what is happening to us as a family while we are growing together, without being limited to what my ideas of family were at any given time. Accepting and building new insight is an active part of growth.

When I began creating the *Bios* pieces, my first child was a toddler; so, the focus of the work centred on the concept of personal growth. A stronger focus on physical transformation will probably become more apparent later on in the series as the viewer will become more aware of the kids growing and the parents ageing. But then there is the idea of a more conceptual growth within the work – the growth that comes with the learning acquired by being a mother and a partner. Before becoming a parent, I had clear ideas about the ways in which I would parent my children. Growing up, I suffered from the lack of interaction and communication that my parents had with each other and with my sister and I. They were good and loving parents, but, blame it on the generation and their personality, they missed so many valuable moments for offering support and guidance. And although I started digging into these memories only recently, I have always felt something was missing in our relationship. That is why I wanted to offer my children a different, more liberal environment where they could express themselves and be who they were going to be freely – something that is more easily said than put into practice.

Today, children are the focus of the family and lead on decision-making much more than in earlier generations. Therefore, children require stronger guidance for their rapidly growing emotions and responsibilities. Children have voices that want to be heard and, as parents, we need to let them be heard, but also filter, buffer, and guide them. When my own young children started showing their personalities, tempers, and opinions, I struggled to accept their independence – especially when it contrasted with my own expectations of them. In my youth, I always tried to stay away from conflict as I felt ill-equipped to manage such situations. But conflicts are unavoidable, especially with children growing up together. Although it is clear to me that I should be a role model for my children and show them how to accept and manage their strong emotions, the ideal environment I ended up wanting for my children became similar to the one I had. So here came a different kind of growth – a growth that begins as an adult, one that I was not wired for, one that involves a loosening of control to embrace the unpredictable.

In Bios 2015, the mould is growing above my eyes and my throat in the self-portrait piece. It is a clear acceptance of my personal growth at this adult stage, as a mother and a wife, but also as an artist. Recognising both the work we have done, and the work that still needs to be done, to create a balanced life is the first step towards a more grown-up state of mind. You could say that a balanced life is a form of utopia (I could not agree more), but there is such value in the journey to achieve it that it does not really matter anymore whether that perfect balance is achieved in the end or not.

Author Biography

Roberta Trentin is an artist and photographer working in New York City. She produces self-portraits, family portraits, and experimental images exploring concepts of growth. Roberta earned a master's degree in Agriculture Science and Technologies from the University of Florence, before moving to the US where she obtained certification in Photography from the International Centre of Photography in New York City, completing a Director Fellowship and participating in their 2012 and 2013 exhibitions. Her ongoing project *Bios/βίος* has been awarded the 2013 Grand Prize from NYC4P (the New York Center for Photographic Art) and First Prize from the Ministry of Science, ICT, and Future Planning in Seoul, South Korea. Her work has also been exhibited in Shanghai and in galleries across the US. Her work recently featured in Musée magazine in an issue dedicated to photographic culture in science and technology. More on her work can be found at http://trentinroberta.com/.

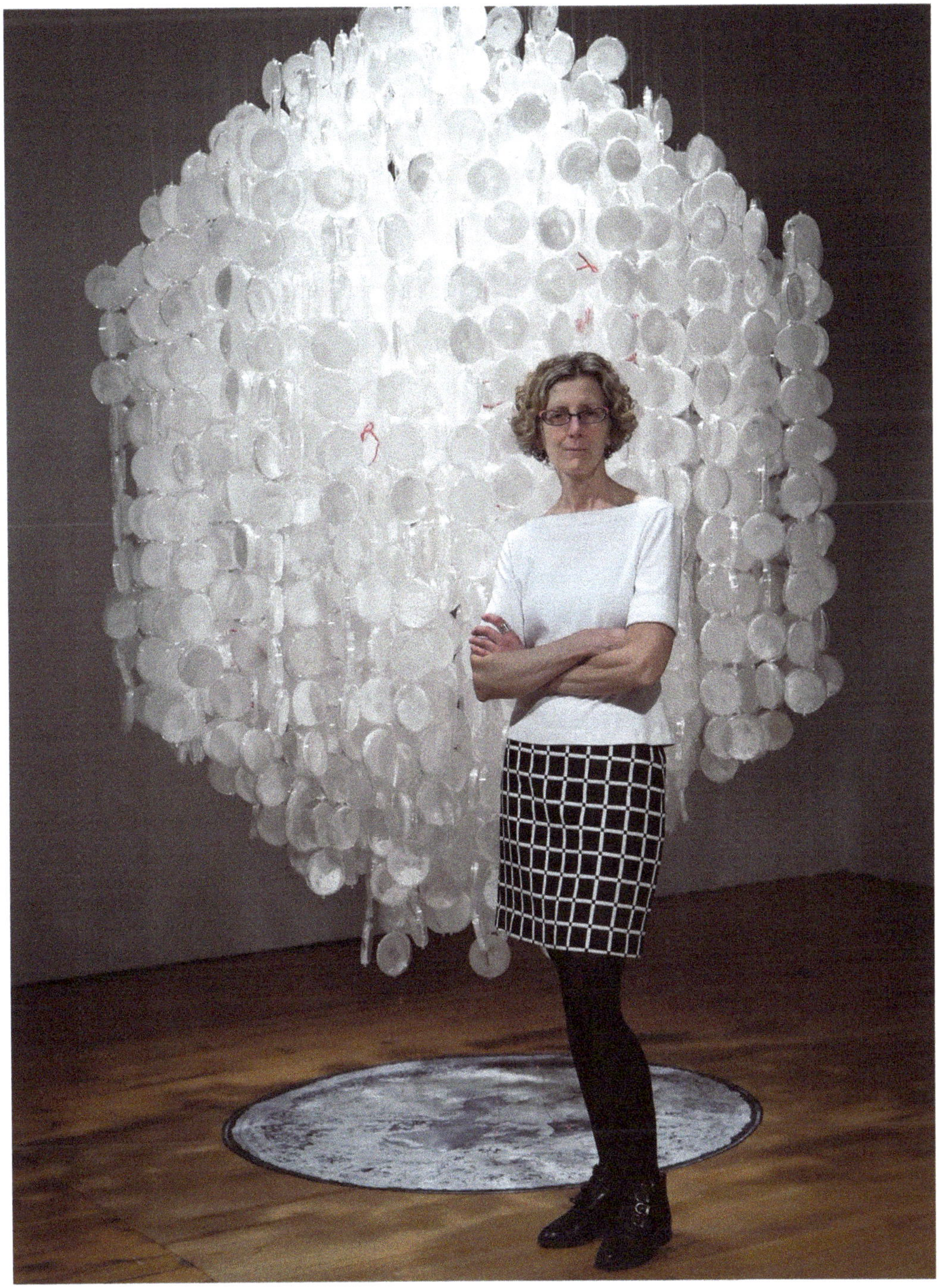

Elaine Whittaker, *Shiver* (2015).

Living in a Porous World

Elaine Whittaker

Whilst microbial life is an essential component of life on earth, it is, none the less, a source of threat to human health. As porous bodies, we can be breached and infected – our body integrity challenged. Are we living in a time of heightened social anxiety around microbial life? In this interview, Elaine Whittaker explores the enormous power bioparanoia can have over us.

A core focus of your work is the constitution of healthy and threatened ecologies. Micro organisms not only pose a risk to human health but also play a key role in making healthy human life possible. How does the tension between these two forces play out in your work, especially in light of escalating social anxiety around microbial life?

The tension between microbes that keep us healthy and those that cause infection and disease is a source of immense intrigue to me. It is at the core of most of my current artwork: How can we connect an artistic aesthetic with the microbes that threaten our fragility as porous human bodies living with organisms that move easily across any boundaries. Ed Yong writes in *The Atlantic* that, 'It's also clear that they (microbes) play vital parts in our lives, calibrating our immune systems, digesting our food, protecting us from disease, influencing the effectiveness of our medicines, and perhaps even affecting our behaviour' (1). Yet, even with all the necessary roles that microbes play, they elicit a great paranoia in each of us individually and in popular culture in general. Even though we are individual, and social, living ecologies (a thriving community of organisms), our relationship with microorganisms remains strained. This strain is linked with unease and paranoia, constant companions that nourish a fascination with the power of the tiniest of life-forms to cause disruption – all themes at the heart of my art practice.

In his remarkable book 'A Journal of the Plague Year' (written in 1722), Daniel Defoe integrated historical records and his own personal experience to recount the 'dreadful visitation ' of the bubonic plague as it swept across London in 1665. Such visitations of microbial plague are hardly a thing of the past: Leprosy, malaria, tuberculosis, the plague, and others, still haunt the world today. The acceleration of global climate change is facilitating new outbreaks of these older menaces, but also giving rise to the germination of new infectious diseases. As a consequence of two centuries of effluent emissions (in the form of carbon dioxide, methane, nitrous oxide, and other gases), climate change contributes to a massive increase in the quantities of hosts and vectors of pathogenic microbes. Seen through this lens and the [re]emergence of old and new infectious diseases, my artworks deliberately challenge the viewer to confront both their own personal, and wider societal, fragility amidst renewed microbial scourges: We live in a porous world, and in porous bodies; the possibility of being breached, infected, and losing body integrity is always present. Viewers are invited – even forced – to consider the cellular communities constantly transforming our bodies and our social ecologies. Microbes are not merely 'visitors' and we are not merely hosts; we are intertwined, symbiotic, and fragile together. My artworks explore the aesthetics of tension – of disaster – and of the unknown. After encountering them, viewers often leave slightly unsettled.

Far from trying to assuage a world threatened by infection, your work instead explores the interplay between terror and beauty, a moment when the scale of a threat transitions us from fear to an admiration of the grandeur of the threat. How does this `aesthetics of disaster' play out in your work?

The possibility of contracting an infectious disease was palpable for everyone with the threat of SARS in 2003, Ebola in 2014 and 2015, and more recently with Zika and Lyme disease. The media, especially social media, highlighted certain aspects of these epidemics in a way that heightened peoples' fear. But the complexities of these infections seldom came across, nor the politics of the various countries involved, the economic structures in place that hinder transparency, and the cultural traditions of local groups that impact disease spread, and so on. Even if this knowledge had been available, the heightened public awareness – one of tension and dread – escalated forms of fear, and often also racism, that intimately associated these diseases with a fear of 'the other'. With the SARS threat, people started wearing masks on public transit, and coughing was looked upon with great suspicion. In North America, we also saw this type of behaviour with the Ebola threat, when intense screening at airports, borders, and hospitals came into common practice. I experienced this screening myself upon returning to Canada from abroad, and once again when I went to a local hospital for a hearing test (here, I was subject to answering a barrage of questions while standing behind a painted line on the floor before entering into the hospital). I realised this was an extraordinary time that begged for an artistic response. Interestingly, in developing that response – the installation 'Shiver' – I was able to alleviate my own personal fear responses. It is always a challenge to move past an initial fear to a more constructive and managed reaction.

'Shiver ' is a multimedia artwork intended to steadily seduce the viewer to move away from a fear of the viral, of the microbial, and even from media scares of impending pandemics, and towards an appreciation of the paradoxical beauty present in the microorganisms that live in, on, and around us. There are two central components to the installation, one sculptural and the other photographic. Though different in terms of media and material, they both present the possibility of seeing the unseen world of pathogenic microbes. The sculptural piece (also entitled 'Shiver') is a 'mutating organism' composed of pipette tips and over 2 300 petri dishes containing home-grown salt crystals; hung from the ceiling, it appears to erupt out of the confines of a petri dish. Some of the dishes also contain salted red wool strands shaped like the filamentous Ebola – described by Frederick A. Murphy, a virologist at CDC (the US's health protection agency), as a 'dark beauty – [a] horror ' (2). 'Shiver' becomes both an object of grandeur and threat as white salt crystals radiate light and shimmer like a chandelier. This is especially effective when the work becomes 'alive', stirred into movement by viewers passing close by and causing air currents to move through it. The second part of the installation – 'Screened For' – consists of a series of photographs of myself wearing medical masks painted with an array of microbial infectious diseases, as found in microscopy illustrations in medical texts. With my eyes closed, or tentatively peering out, these enlarged life portraits are disconcerting and eerie, yet also purposely beautiful – tightly cropped photographs with inviting sky blue masks depicting large colourfully painted microbes.

Your practice has come to encompass a wide range of media, from home-grown crystalline forms to the Archaea `Halobacterium'. In overlaying cultural imagery with living materials on the subject of infection in your work, you are able to generate a deeply layered engagement with your subject. How important (or necessary) is this interplay in your work?

Combining cultural imagery with live organisms has been essential to many of my artworks. The images I have selected from popular culture have made the works richer material-wise and encourage a more complex and symbolic interpretation. The viewer is forced to make comparisons between memories and associations triggered by a cultural image and the 'live organisms' that obscure and scar that image. This technique is employed in my installation 'I Caught it at The Movies'. I believe most viewers think of themselves as impervious to the gravity of infections of the sort that run amok in pandemic movies. Even if they recognise that pandemics are always a possibility, the comforts of western capitalism seem to make it appear as something highly improbable. The viewer recognises that these are movies, that the people depicted in these films are only actors, and that there are no real victims or survivors of these 'infections'. They recognise that these outbreak narratives are popularised references transmitted through the medium of film. Oddly, they gaze at scientific depictions of bacteria and viruses with hypnotic admiration and terrified awe. This reaction is no different from staring at the countless visualisations of viruses and bacteria that grace the covers of magazines and journals: They too both attract and disgust us with their luscious colours and implied-potential for catastrophic effects.

In 'I Caught it at The Movies', there are hundreds of the petri dishes containing stills from a wide range of movies from diverse genres (such as disaster, science fiction, horror, etc .) overlaid with painted microscopic visualisations of infectious diseases, themselves enhanced by halobacteria. Episodes of confrontation take place that are culturally, historically, and scientifically charged by the imagery, and the viewer is forced into self-reflection about the nature of infectious diseases and their possible impacts and consequences. The intersecting motifs of cinematic and popular memories positioned as scientific objects inside petri dishes are constituted for close examination as if ready for the microscope, guiding the viewer into a world that collapses fiction and reality: What is real and what is manufactured? These probing, artistic engagements are meant to destabilise and unnerve the viewer, just as our own daily experiences in the world are becoming increasingly tenuous and disrupted.

Of all my works I have produced to date, it is this installation that has received the most public reaction. People identify with the disaster movies I have selected, so, without even the need to show gruesome details, they were compelled to get as close as possible to the bacteria-laden dishes to identify which movies or celebrities I had inserted into them. In some cases, audiences even made a game of linking the movie with the outbreak – rabies in this one, plague in that one, and so on. Through this more intimate encounter, many expressed their surprise at finding the colour and crystallisations of the bacteria in the works (and the agar on which they were grown) as quite beautiful. This created the opportunity for further discussion about the social anxiety and terror associated with microbes. On further reflection, it also forced the viewer to consider the unavoidable role of bacteria in our lives, and that we cannot possibly insulate ourselves from our natural ecology. The installation effectively transforms Hollywood 'shock and disease' movies into a radical aesthetic gesture of quite a different meaning.

In one understanding of media, your work opens a dialogue with your audience, engaging with issues around fear and infection or the grandeur and fragility of the human body. The use of halobacteria, however, introduces a perceived, real-time threat of infection into this conceptual space. Beyond conversation and reflection, how do your audiences react to the living material you present?

The cultured bacteria in my artworks generate viewer reactions in almost every way you can imagine – from intense interest and a need for closer inspection, to astonishment, surprise, trepidation, and even dread and fear. For some, it was a 'wow' moment, drawing them to lean in closer for a more intimate visual experience. For others, it was unsettling just being in the gallery with the bacteria: They were visibly uncomfortable, asking for assurance that everything was safely contained while keeping their distance, not venturing too close to the works. And then there are always a handful of viewers who are repelled: They would start by reading the artist statement, and, as revulsion and fear spreads across their face, they would scurry out the door before even looking at the works. I must say that all these reactions are the kinds of engagement I expect – I even get it from friends visiting my studio. Confronting the social conditioning around bacteria is one of the functions that has to come from 'new art' revealing the unknown through an aesthetic practice. For others, it opens an entirely new way of thinking about social and political issues today – and that is even more satisfying. The varied reactions of viewers could be looked on as an interesting gauge of our times. Is their fear even greater now because we are more aware of infectious diseases and how they spread? Sensationalised news reaches us with even greater speed, even though it might not always be factually true. It is hard to say how this impacts the dialogue between the art and the viewer. But because these minute life-forms are so closely associated with fear, the use of any bacteria in my artworks, even if it is non-pathogenic, is anxiety-inducing.

Brian Massumi has discussed how infections acquire real status not through an effective occurrence but through anticipation: `The resulting fear becomes rather pervasive, since it is a reaction to a `quasi-cause' that hasn't manifested yet, but might (or might not) occur at some point in time' (3). That anticipation is subject to an always-incomplete picture of scientific, social, historical, and personalised knowledge. Is this where the richest exploration of fear lies?

There is something very real in what Mike Davis referred to as an 'ecology of fear' that extends from policy and surveillance to the bio-panics created by public health warnings. We are intertwined in this ecology of fear, in the ways we live now, even if the different components of that ecology have distinct causes and suitable responses to them. Thinking back to the disaster movies, filmgoers might leave a cinema either buoyed by characters who have survived against all odds or burdened with a simple dread because the threat of infection 'remains imminent'. In a gallery space, in comparison, visual art can offer something more than a simple narrative encounter with these issues. I try to create the anticipation of an unpredictable threat in my works as a kind of tension – installations that are beautiful and enticing, but also unsettling to the viewer as they enter a world of bacteria, mosquitoes, and disease. This individualised ecology of fear is confronted right in the gallery: It is an acute reaction (fear playing out biochemically in sweating, increased heart rate, etc .) coupled with individualised contagion anxiety. I present these artworks not to explain a general social ecology of fear but to reveal some of the tensions that exist between the beauty of microorganisms and the ever-present fear of their hidden danger.

In this sense, raising awareness on the issues of disease risk and its management is not the main intention of my work: I do not purposely make work to be didactic. My artworks are, however, scientifically, socially, and historically oriented. A fortuitous consequence of this is that they are able to generate discussion and debate about such things as disease awareness. Often, my works are based upon extensive research in medical and scientific documents or on events from history. It is, therefore, appropriate that they are exhibited in science centres, in science –art galleries, and even at medical conferences. Such venues are not primarily art spaces, but they do foster in viewers an expansive interpretation of the ideas and concepts I explore in my work (turning them, although unintentionally, into educational discussion points for visitors, particularly children). The works also force more reflective moments for scientists and medical practitioners, sometimes taking them onto unexpected trains of thought. Even in more typical art settings, the engagement of viewers with my work does not centre on 'the facts' behind infectious diseases or on how to manage the threat of disease but, rather, drives an aesthetic impact that may trigger discussion or provoke something more unsettling.

Many artists who work with living media today had beginnings in science education or research (or a combination of both). What was your own entry into this field and how do you see it developing as your engagement with it diversifies?

I do not have a formal science education. Instead, my early artistic practice began with photography and sculptural ceramics and overlapped with social activism in the ecological movement and around feminist health issues. I formalised my education in art much later but continued to be involved in social justice issues, meanwhile pursuing an art practice that included untraditional materials such as wax, insect bodies, and salt. The driver for incorporating organic materials into my art comes from a fascination with the corporeal ecology of the body, with medicine, and the natural environment we are embedded in. I started portraying these notions first through salt: by working with salt, a mineral, I was able to mimic the organic, growing and nurturing diaphanous crystals on created and found objects. Many viewers perceived these crystals as organic because they were grown, but they are, in fact, lithic, geological, and inorganic – a mineral, not a cell. I was also drawn to salt because it is the foundation for life, a link from our primordial past in a briny ocean to our foetal beginnings in the salty milk of amniotic fluid. It is also the most common inorganic substance in the human body. Trespassing the boundaries between organic and inorganic (and between the microscopic and macroscopic) salt became both my main material and metaphor in my early artworks. Alongside salt, other materials such as wax, bone, mosquitoes, and plant organics became an integral part of my material repertoire.

It was when I was researching the history of pandemics, early microbial life on earth, and the rise of infectious disease under the ecological pressures of global warming that I was moved to incorporate live bacteria into my artwork. When I learned that there was a non-pathogenic salt bacteria that could be easily obtained from a biological-supply company and cultured in my studio, I realised I had found the metaphorical and material stand-in for the infectious diseases I was studying. Having live organisms in my work coupled with cultural and political indicators made for a fuller, enlivened, and challenging experience for the viewer. What often begins as just another afternoon at the art gallery soon dissolves into a very different confrontation with artworks that become disconcerting ideas about our personal and social ecologies.

The cooperation between professional research scientists and artistic practitioners is becoming more established as a means of creating innovative work or exploring the porous boundaries emerging between traditional disciplines. Do you feel artists could, or should, have a place in the research lab of the future?

It is true – and still surprising to me – that more traditional scientific laboratories now support (and even fund) artists to become collaborators. The idea of what constitutes a lab is, at the same time, becoming incredibly complex in terms of its engagement with interdisciplinary thinking. Artists and scientists are being brought together under novel 'umbrella' structures – such as science galleries, science centres, ecological organisations, and specialist institutions in astronomy, particle physics, nanotechnology, biotechnology, and more. I have had the opportunity to exhibit in a number of these impressive venues and have also worked with boundary-bending scientists on projects exploring, for example, dance and biology, toxicology, tissue engineering, and augmented biology. These quite surprising collaborations are likely to continue evolving in unexpected ways in science laboratories. This may, indeed, be driven by the tremendous – even frightening – fragility of our current political and ecological systems.

Paradoxically, the worse things get, the more there is for art-making to reflect upon.

Perhaps with the expansion of living media practices, we can start pointing towards alternative futures than those that seem likely at present? That might be far off though; artists working with living materials still face a number of challenges in getting their work seen by the public. This is not surprising, given the very limited art market for such work. The transitory nature of these pieces and the expensive upkeep of live organisms make it even more challenging for individuals and institutions to display work. Funding for our art is slowly getting better, but the fact is that many jurying members of funding organisations either do no t understand the science or our aesthetic sensibilities. Even transporting biological artworks across borders can be difficult as security concerns can hold up or even embargo the work indefinitely – as recently happened to me. So, supportive organisations that bring together artists and scientists (like the ArtSci Salon in Toronto and the SciArt Center in NYC) are crucial for artists, like myself, to aesthetically respond to future contagions. Who would have thought over two decades ago artists would be making artworks with microorganisms?

Working with living materials depends on tools that have, in their time, revolutionised Science and become mainstays of scientific practice. Are the techniques you use stabilising, or is this still a ripe period for invention and advancement in the arts?

This is an area of great inventiveness within arts practice, in part thanks to a wealth of cross- fertilisations between scientific work and artistic activities. For example, new techniques and processes in the lab are being developed and keep evolving (such as CRISPR – a process that helps make specific changes to DNA in plants, animals, and humans; a technique now being taken up by artists). New biotechnologies are also being taken up by bio-hackers in the DIYbio movement to transform living forms, including their own bodies. The consequences of this intersection between science and art-making are often a significant transformation of traditional approaches in painting, sculpture, and drawing. There are whole new ways in which the boundaries of art are being redefined today. As Artist-in-Residence at the Pelling Laboratory for Augmented Biology in Ottawa, my own work is taking an unexpected path as I learn about the process of decellularisation in living and vegetable matter. The Pelling Lab is known for carving an ear out of an apple and culturing it with mammalian cells. By removing from apple cells and DNA, only the cellulose scaffold (that gives the apple its structure) is left. This artificial scaffold can be reshaped and implanted with stem cells to potentially grow replacement organs and tissues (such as ears). I have been collaborating with them to use this technique for a new art piece – a decellularised maple leaf cultured with human lung cells. Whilst the lab focuses on the possible future application of this approach in the medical field, my piece (for an upcoming exhibition) is much more speculative. It was envisioned as a metaphor for trees: as lungs of the earth given even greater – even fantastical – potential by harbouring and combining with human lung power. My own practice is being enhanced through exposure to the expertise and experimental processes of laboratory work. Nonetheless, being innovative in my own right as an artist is a central driver for the creation of new work.

References

(1) Yong E. You're Probably Not Mostly Microbes. The Atlantic. 2016 Jan 8 [cited 2020 Sep 25]. Available from: https://www.theatlantic.com/science/archive/2016/01/youre-probably-not-mostly-microbes/423228/

(2) DelViscio J. A Witness to Ebola's Discovery. The New York Times. 2014 Aug 9 [cited 2020 Sep 25]. Available from: https://www.nytimes.com/2014/08/08/science/a-witness-to-ebolas-discovery.html

(3) Massumi B. Fear (The Spectrum Said). *Positions*: asia critique. 2005; 13(1): 35. Available from: doi.org/10.1215/10679847-13-1-31.

Author Biography

Elaine Whittaker is a Canadian multidisciplinary artist working at the intersection of art, medicine, ecology, and biological science. Her art practice principally focuses on the creation of installations, which include sculpture, drawing, painting, and digital imagery elements. Her work has been exhibited in galleries and museums both nationally and internationally, including in Canada, Mexico, France, Italy, the UK, Ireland, China, South Korea, Australia, and the US. Her work has also been featured in digital galleries, literary, medical, and art publications (such as William Myers's book 'Bio Art: Altered Realities' (2015). In 2018, Elaine was Artist-in-Residence at the Ontario Science Centre and is currently Artist-in-Residence with the Pelling Laboratory for Augmented Biology (University of Ottawa). More on her work can be found at https://www.elainewhittaker.ca/.

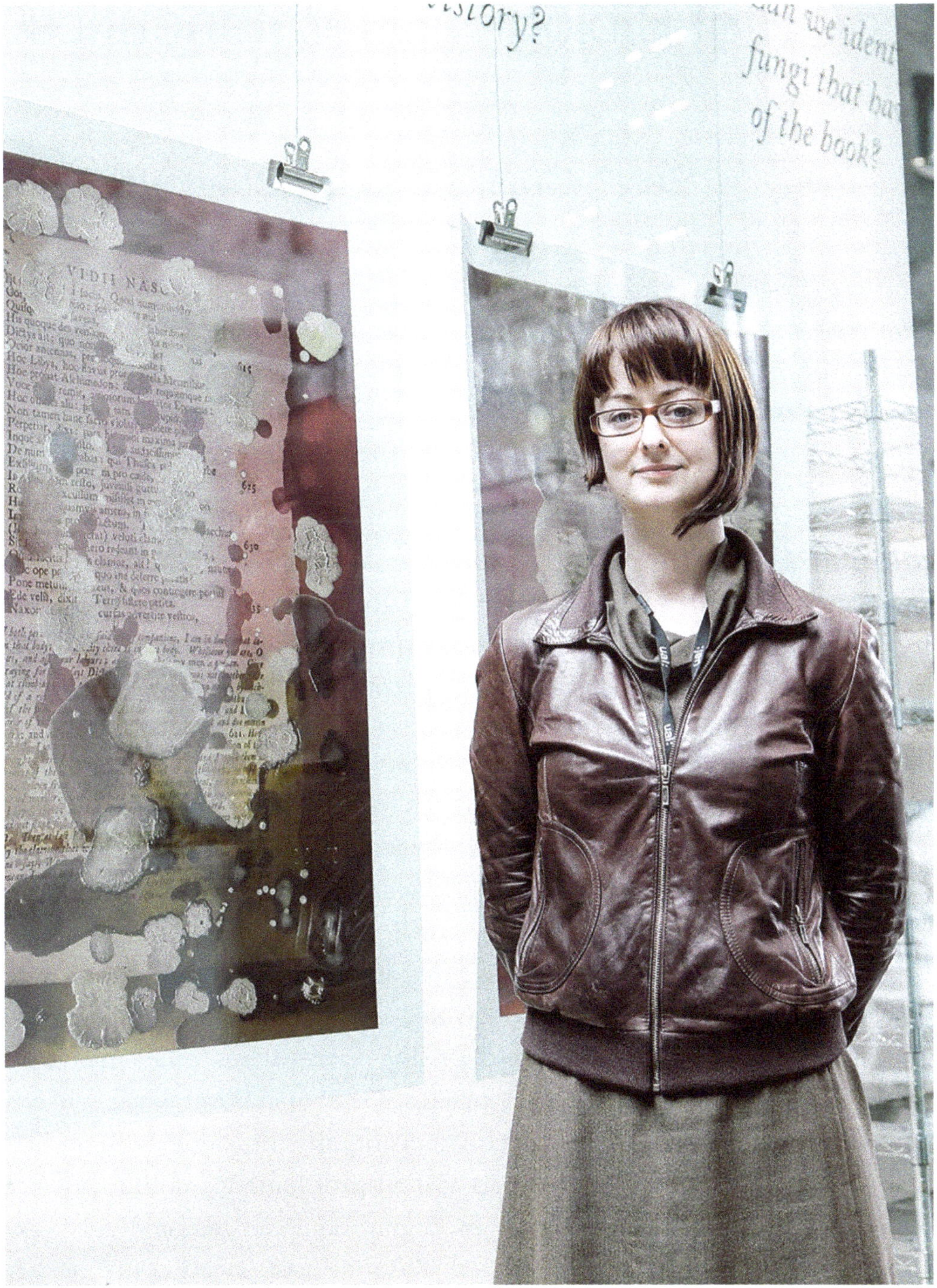

Sarah Craske, *Biological Hermeneutics* (2017).

The Art of Biological Hermeneutics

Sarah Craske with Dr. Charlotte Sleigh

Our literary archives are a record of our written culture, revealing what we value most to future generations. For Sarah Craske, it is not their textual content but their support of microbial life that opens up new ways to read an archive. In this interview, Sarah asks how microbial life challenges our understanding of archival practice and the institutions that keep it in place.

We are accustomed to think of physical archives as stable repositories for knowledge but their future seems increasingly under threat. How are archives changing today, and how does your work help us think about that process differently?

SC: Over the last decade, libraries and archives have gone through a huge process of change. As technology continues to develop, so does our relationship with knowledge. In consequence, the status of knowledge, and access to it, is continually being redefined: Knowledge acquisition and storage is moving from the real to the virtual world. Libraries, as interdisciplinary research centres, act as both an archive of knowledge artefacts and as a digital information highway. Both roles are being expanded, but also merged, over time as a focus on digitising archival materials increases. The expansion of digital material prompts the question: What will our relationship with the physical archive eventually become? Will it hold any value at all? Digital archiving is not unproblematic; unlike physical artefacts that can survive for centuries, digital data can become corrupted and digital data formats soon outdated. Through fast-paced consumerism, even the equipment needed to access data created a decade ago is becoming rarefied. A dystopian prophet might predict a digital world where physical objects are no longer conserved and safely stored but, rather, discarded and scattered across a landscape – their perceived value lost once digitally appropriated.

'Biological Hermeneutics' is an artwork I have developed in collaboration with Dr. Charlotte Sleigh, Dr. Simon Park, and Chethams' Library in Manchester to interrogate this situation further. As a project, it essentially reveals a transdiscipline – one that questions the tension between these digital and physical states through enquiry that crosses scientific and artistic boundaries. Through it have emerged new theory, writing, and a variety of mixed media exhibitions and performances. The premise of Biological Hermeneutics is that the physical archive is not merely made up of written or printed text; it also contains data embedded within its biological forms that reflect its usage and those who use them – books as centres of microbial data and data transfer that forces us to question how we interpret texts and the means by which we interpret them. In this way, we are asking whether a move from traditional conceptions of archival taxonomy and practice might be possible, so opening up forms of archival knowledge and understanding that might not even be conceivable at present. In the background to this work is a deeper acknowledgement of society's imperative need to move from an object-based, commercial, and material-use culture to a sustainable, ecologically concerned, objectless culture. As such, it reflects on the 'death of the object' in art history, in museology, and, quite literally, the process of decay in physical and digital objects.

How does the concept of Biological Hermeneutics bring together the different scientific and humanistic methods required to form a new understanding of the archive in this way?

CS: Hermeneutics concerns the process of interpreting texts and probing the ineliminable gap between author and reader. Hermeneutics came to fame as a method of biblical study in the early-to-mid 19th century through its questioning of the many assumptions underlying biblical literalism. We see here that early critical questioning of Christianity came initially from literary methodologies, not scientific methods as is commonly imagined. We want to understand how hermeneutics can be performed in relation to something assumed to be open to scientific research but closed to humanist study – in this case, biological life.

SC: Biological Hermeneutics is, then, a form of study in which books are read for their biological, rather than textual, content, such as the microbial life (e.g., bacteria or viral) on their pages. Our use of the phrase is deliberately ambiguous in grammatical terms: It can mean doing hermeneutics through biology, or doing biology through hermeneutics. This strikes a connection with natural philosophy ; this is particularly important since this early modern form of scientific practice included the kind of philosophical inquiry towards meaning that is often excluded from the Sciences today. There is, therefore, an important symmetry here of approaching biological subject matter through humanities-style methods on the one hand, and looking at a text through scientific methods on the other.

CS: There has been a trend for the latter more recently, for example, in new concepts such as neurocriticism or neuroarthistory. Here, there is a risk that a particular type of knowledge – the correspondence between brain activity and human behaviour, such as when enjoying a poem – is seen as trumping or subsuming humanist knowledge: Even if we can detect through a brain scan that a reader has a strong response to a poem, it is not the case that we no longer need to spend time elaborating its interpretation. In my opinion, these two angles of approach (from scientific and arts/humanities methods) must be held in constant tension, without allowing one to overcome the other through disciplinary bias.

As part of this new transdiscipline, how have you gone about uncovering the microbial life of archival materials?

SC: We have developed what we call the Biological Hermeneutic Print. Through it, we are able to isolate and recover viable bacteria from the pages of a book. This is done in a way that preserves their spatial relationships on the surface of the page itself.

First, a 'Molten blood' agar base (a highly nutritious general-purpose agar) was poured into bioassay dishes and allowed to set. This is a type of agar used to grow organisms with complex (termed fastidious) nutritional requirements; it is highly suited to reviving damaged bacterial cells from very old books. Using aseptic technique, we pressed pages from Book Three of a 1735 copy of Ovid's Metamorphoses onto the agar plate surface we had prepared. After 20 seconds, the pages were removed and the plates incubated at 25°C for at least a week to encourage bacterial growth. Where bacterial cells were transferred from the book onto the agar, they multiplied over time to form visible colonies (now containing billions of bacterial cells). In order to allow colonies to fully develop, they were left to form over a period of many months. A wonderful twist in our method is that not only microorganisms but the paper indentations arising from the original letterpress printing some 300 years earlier were captured on the agar's surface – maintaining something of the relationship between a text's biological and symbolic content.

Once developed, we were able to isolate colonies from our prints and use a technique called dilution streaking to generate new colonies (subcultures) with pure bacterial strains ready for DNA analysis. Here we found a type of bacteria that only exists 20,000 feet into the Earth's atmosphere, which can only be explained by the flight I took to New York with the book in my luggage! (The ways in which we are unexpectedly complicit in the life of the archive emerges once more.) We also learned something of how readers from the past have physically interacted with the text itself. This has included finding more bacteria common to the human skin on the Latin rather than English translation of the text. On one of the pages, a sneeze could also be evidenced, both in how the bacteria had been dispersed across the page and the type of bacteria identified (one commonly found in nasal passages). Rather than ending our research here and disposing of the used agar plates, we carefully dried the agar to generate a thin glass-like film that fixes and preserves the bacterial life on their surface. I then designed and built an archive ready to store the bacteria harvested from the book. After DNA analysis, the samples were then labelled and stored in the archive, so creating our own 'microbial library' – the first expression of our Biological Hermeneutic transdiscipline. We will continue to update the library with more samples as the opportunity arises.

By uncovering the microbial life of archives, you reveal forms of living process and exchange that have been sustained over very long periods of time – even over centuries. How would you characterise this ecology (or these ecologies) long-hidden from view?

SC: History, as is implied in the word itself (from a western perspective) is the *written word* – stories constructed with intent. Unknowingly, or without consideration, we have created agents to collect alternative and unmediated histories – from the microbial life of those who have constructed the texts to collections of environmental data that can offer unmediated socio-political narratives. Each ecological community is also singular, which becomes apparent when you compare different archives: All have their own unique profile depending on the focus of the archive and, for example, the demographics of its users. Archives can, controversially, be profiled according to their unique composition of microorganisms.

CS: The humanities are at an interesting moment in their development. For three hundred years, we have been the inheritors of human exceptionalism, an exceptionalism that asserts our view (sometimes aggressively, sometimes resignedly) as the only view that is obtainable to us. In other words, this is a perspective that leaves no room for a God's-eye view and no form of objective view from outside of our own selves. The philosopher Kant proved that we cannot grasp the fundamentals of the cosmos (the noumena) apart from through the limitations of our human senses and experiences. But now we see post-humanism challenging this: We are starting to talk about how 'critters' (Donna Haraway) and 'things' (Bruno Latour) have lives of their own. As a counter to human-centred hubris, this seems like a good thing. However, the Kantian question remains: Can we actually see, take account of, and interact with these *other things in the world* apart from through our human goggles? In other words, can we ever relate to them – through knowledge, through relationships – in their own right? We do not know the answer to this question, but the perspective of giving archival microbiota space to grow and interact on their own terms (outside of our control) seems a beautiful and mysterious meditation upon this, perhaps, unanswerable question.

With the concept of Biological Hermeneutics, you make explicit the web of human interactions and reasons-for-action that are the life of an archive. How does this work reveal connections between different elements of cultural, social, economic, and natural history that constitute the archive?

CS: The archive is deeply constituted by connections between these different elements. This is why *the archive* is a problematic term in that it materialises and canonises assumptions of value – the culture that matters – in a way that has, historically, reflected various skews on gender, race, ability, sexuality, and so on. But what we have enjoyed discovering is a little bit more about how our bodies, too, are an archive of microbiota from deep and recent times. This radicalises and opens up the archive – and often in very problematic ways. With the concept of Biological Hermeneutics, we are interested in questions of how representation and reality can convolve [1], i.e., how the map of the landscape becomes the landscape itself or how the archive of the world becomes, in itself, the world (a very Borgesian idea). This is more easily illustrated with an example from Sarah's work – one in which the institution involved has to remain anonymous.

SC: Whilst Artist-in -Residence, I implemented a couple of public engagement events, one of which specifically targeted the staff at the site (numbering over 100 people). The research activity (built around the 'microbio ta collection trolley') involved was very simple: They were asked if they would voluntarily impress their fingers onto an agar plate; the samples were then fully anonymised but linked to information on which parts of the site, and specifically which archives, they had visited. Subsequently, the microbiota samples were nurtured on agar plates and their growth documented photographically. Through visual results alone, clear patterns became visible. You could, for example, identify which archives had been visited based on the visual similarity of microbiota growth patterns, i.e., a group of people who worked more exclusively on one archive had similar microbiome characteristics – perhaps even convergently so – through interacting with common objects. Further, the results seemed to suggest trends that matched the hierarchy of staff: Those higher up the status ladder had less microbiome diversity than those further down the ladder. You could also start to identify teams: The personal assistant of one member of staff shared similar results to the person they were assisting. This work highlighted relationships and interactions within the archive that had not really been questioned before. The exchange with microbial life through interaction with archival materials has always been present (and will continue), only it was not apparent to our human senses. In an institutional setting that strives for stability and consistency, this uncovering of a hidden truth can be unsettling.

[1] We are indebted to Romén Reyes-Peschl for introducing this term, which is of his own devising.

If we become aware of the microbial life picked-up, left behind, or transmitted to others through interaction with the archive, is there the possibility for new practices to emerge that use the archive as a targeted means of a symbolic cultural and intellectual exchange?

When the question was raised as to whether individuals, or an individual's socio-economic status, could be identified from the microbiota samples – *a question I simply could not answer* – the results of the project became highly controversial, causing extensive debate within the institution. Their ethics committee became involved, and, finally, over fears that the work might be breaching data privacy and intellectual property law (as each sample is a unique reflection, indeed creation, of an individual), the project's results were pulled and I was restricted in what I could publish as an artist. There were questions about the legal status of individuals, the social function of archives, and the ethics of institutions that the department was not ready to 'deal with'. What the operation of the human microbiome means for these issues is genuinely an open question.

It is interesting to me that, rather than embracing the potential for exploring this exchange between the physical archive and archivists, such defensive measures were taken within the institution. I choose to work with archives in this manner precisely because it is so different from the flexible interdisciplinary space I work in. Therefore, I always acknowledge the risk that something unpredictable might be uncovered in my work that creates tension with a knowledge-stabilising mentality of the institution. I had hoped that new approaches to cultural and intellectual exchange with archival life would be possible, but it might just be too soon. Instead, I can certainly envisage policies being written in the future to *manage* archival microbiota, with new value systems put in place to navigate the *risk* it presents. Unfortunately, like any resource, I can only see it being mediated and exploited to some end other than a potentially exciting, and highly original, form of cultural exchange.

CS: What that alternative might look like is anyone's guess. The archive is already, symbolically speaking, a heavily freighted object of cultural value and exchange. Perhaps thinking of their microbial content could drive a democratising move – a spur to broadening who gets to participate in archival life? This also puts me in mind of faecal microbio ta transplants (a means of transferring critically important bacteria from healthy individuals to the guts of individuals who are health-impaired); perhaps, we should be thinking of scholarship along these lines too – a form of microbiota transfer that brings us closer to the lives (and agency) of other scholars and scholarly communities. That would certainly up-end a few things!

With this new outlook on what constitutes a material archive, archival practice, and the interactions that give an archive life, do you now think about the cultural movement to digitise archives differently?

SC: Once we recognise the central issues around how data is essentially mediated (who chooses which data is to be shared, in what formats, how it is edited, what is valued in that data, etc .), the digitisation movement becomes so much more problematic. With over a decade's experience of working in libraries, it seems clear to me now that the digitisation of knowledge is directly linked to neoliberal capitalist principles that have a different value framework to those of the archives themselves. Libraries are constantly having to reduce or manage their physical archives due to the demands they make on 'space' – which of course has a fiscal value. The short-termism of converting space into income is threatening the life of the physical archive. This is ironic as the storage of data requires enormous server centres across the world, each with their own huge demand on energy. The physical consequences of the digital world are now a topic that needs much more attention.

What I find particularly interesting is that the physical archive – especially vellum and the older print materials – are actually more stable than their digital counterparts. Libraries face a constant battle in how best to preserve their digital archives as technology and data formats continue to be updated. One of the libraries I have worked in was spending significant amounts of money trying to restore historic tape machines and deciding on an appropriate digital policy – a decision that will give them probably around five years breathing space before their hardware/software is no longer able to read that file format. With digital policies constantly updated, there is a question of how much data should be collected and stored at any given time. One library I know scanned thousands of detailed Victorian-era maps in order to dispose of the original materials, only to find that much of that detail had been lost due to their choice of image resolution. Marginalia and text metadata can be just as valuable as the text itself: They can provide insight for translation and the basis for new knowledge. Ironically, whilst these conversations take place in the main library, the special collections, often with centuries-old texts, sit quietly in their climate controlled room, potentially for generations.

Of course, there is a central tenet that digitising information democratises data for everyone. But I think this is a false notion, especially if some are then restricted in their access to the physical archive as a direct result. The risk is a further entrenching of the physical archive in its already-elitist status. With a few rare exceptions, original texts may increasingly end up confined to, what are in effect, private libraries and collections. Here, they will develop ecologies of microbio ta that reflect – maybe even one-day betray – this restriction of who gets to use them. Fundamentally, a digitisation movement that essentially *puts the breaks* on those interactions that make archival microbiota possible – without first asking what the consequences might be for the loss of new types of insight – is a little troubling.

CS: We live in the era of big data, and with that comes a sense that data trumps everything else. There are theorists, like Yuval Harari, who go so far to claim that data is all that matters, that we are obliged to facilitate the exchange of data, and that the material carriers of data – animals, plants, humans – take second place. As a Marxian, and an inheritor of Judeo-Christian notions of embodiment, this is a troubling perspective. Losing a physical library might seem like a minor problem, but we have to ask how this is all part-and-parcel of the same 'commodity fetishism' for data (a Marxian term used deliberately). Here, subjective aspects of economic value are transferred into the objective things (in this case data) at the expense of human relationships that actually give data value.

Any discussion about the archive needs to touch upon concepts of media, both as material and communication. Your work reveals the many ways in which we can become inextricably part of an archive's material form and performance. Should we understand this as a process of `becoming media'?

SC: Yes, absolutely. Media studies are increasingly important to the history of science (e.g., in the work of Jim Secord) as we come to realise that no knowledge can come into existence or operate unless it is mediatised: We need to look at media if we are to understand knowledge discourse. I like the pun of media in our work: Both agar gel and books are media for microorganism growth, but also for the growth of human knowledge. In this instance, then, media is not just mere textual material, as it also includes microbio ta, people (archivists and archive users), and forms of exchange that help shape this wider concept of archival life. It is important to recognise that, prior to my residency, archival staff were completely unaware of how they were contributing to an archive's microbial life and working *in relationship* with it – essentially developing another level of *shared experience* or encounter between human and archive. A microbial conversation was occurring, if you like, and, arguably, this could have consequences for a staff member's biological, mental, and emotional states.

The effects of this are not to be underestimated. We like to think of objects (such as archival materials) as rather unassuming in their own right, with no perceived agency, and of a time and place distant from us in the present. When you start seeing how new forms of archival materiality and the interactions between human and non-human agents are deeply part of the archive in the present, that distance diminishes considerably – the media that is the archive includes us. So, I believe we need to start asking different questions about where boundaries do or do not exist between objects, forms of agency, and so on. Do we need to start thinking about how our *own* agency is actually distributed across living and non-living objects? Should an understanding of our agency *include microbiota* as an inherent component rather than as something *other?* To what degree microbiota have their own agency (and whether there are consequences that follow from it)?

Transdisciplinary project outputs are inevitably the start of new conversations, pointing to lines of inquiry that could fall within more established disciplines or orient work towards new unknowns. How has the tension between different disciplinary demands played out so far?

CS: Fundamentally, we think this work adds an important dimension to our appreciation of old texts by seeing them as a medium for microbial life and exchange. That archival microbiota might be used to detail individual's socio-economic status, literary interests, and social interactions has already been raised, but now we are starting to see how this work pushes other lines of inquiry within the natural sciences (such as the science of microbiology), the social sciences, the arts, and the humanities. It is really only through future work that we are going to make sense of some of the insights so far and work out exactly what questions we should be asking in the future.

SC: Certainly, there is more we can ask about how archival materials contain traces of their own history (including their own microbiological history).

So far, as discussed earlier, we have had the opportunity to examine a 1735 copy of Ovid's metamorphoses for both layers of microbial life and atmospheric mineral deposits. The question of whether archival materials can serve as a record of changing genetic information contained within microbiota over time is a tantalising research question to pursue in the future. In theory, this is possible to study: Books may contain dormant bacteria, extremely long-lived active bacteria, or spores released at the point of a bacterial death. It is entirely feasible, therefore, that the genetic information extracted from these samples could illustrate shifts in evolutionary history. From here, it is not a huge leap to imagine how a historic archive could help in the battle against current microbial-related challenges: The scientists on our team, for example, were keen to find older bacterial DNA from a time when antibiotic resistance had not yet occurred. Our initial project, however, lacked the budget to explore these types of enquiry in any further depth.

To be honest, the enormity of the world that has opened up since I started to study these texts through this different lens has forced me to rethink my own role as an artist; it has meant prioritising those elements of the work most important to me. At the moment, these concern questions about how we reframe the relationship between our individual selves and our microbio ta and how we can develop collaborative frameworks that allow these relationships to be studied in a truly transdisciplinary way. Who knows what new ways might emerge for us to *read the archive*?

What, for you, is the promise of transdisciplinary working, and what are some of the challenges you face in putting it on a more stable foundation?

CS: For me, what all this demonstrates is the potential – but also enormous challenges – of transdisciplinary working, something we have written about in two recent articles (1, 2). Transdisciplinary approaches have in their favour that they can be targeted where needed (e.g., object-focused or problem-orientated) and leave behind a lot of disciplinary baggage. (Although, it goes without saying, in any new formulation of inquiry, traces of its predecessors will remain). They can also be positioned in a world where complex problems are not amenable to mono-disciplinary solutions; in this way, they bear some resemblance to 'post-normal science'. Finally, they carry an implication of democratic engagement – if the right processes can be put in place to support them.

SC: There is, of course, a risk that a transdisciplinary approach might efface valuable disciplinary perspectives and methods of critical questioning that they can offer: This means drawing on different disciplinary perspectives in a productive fashion is rarely easy. On the other hand, transdisciplinary working can highlight (as I have seen all too often) priorities stemming from disciplinary culture that then require workarounds in order to remain true to a project's core values. This problem-orientated approach also excludes a whole realm of knowledge creation, namely science or art for their own sake – something which I still believe has an important place in the world.

CS: How we construct transdisciplinarity in a neoliberal culture that values only economically calculable solutions is another important question. Moreover, transdisciplinarity must actively address the charge that can be levelled at any form of investigation potentially *tainted* by its historical siting in spaces of privilege: To do, for example, a form of 'pure science' requires first having no worries about your own rudimentary health, living conditions, income, etc . Transdisciplinary working offers the promise of taking this into account through opening up participation and challenging such issues head-on. For me, the figure of the 17th century natural philosopher Robert Hooke making microscope drawings of tiny organisms found on the pages of his own books is an irresistible image: Recuperating the richness, the epistemological quirkiness, and the phenomenological experience of Hooke's approach to science reminds us of what transdisciplinary working might also achieve today.

Author Biographies

Sarah Craske is an artist working at the intersection of Art, Science, and Technology. She works with multi disciplinary spaces and organisations to drive new transdisciplinary collaborative partnerships. Her work interrogates a range of practices: research, writing, installation, film, performance, sculpture, casting, engraving and printing, synthetic biology, and architecture. Sarah graduated in 2016 from Central Saint Martins in London with an MA degree in 'Art & Science', receiving a distinction and the NOVA award. She was shortlisted for the 'John Ruskin Prize – Agent of Change' in 2019. Recent solo exhibitions include 'Biological Hermeneutics' at the Chethams Library in Manchester, and 'THERIAK – The Past In The Present' at The Pharmacy Museum, University of Basel. Sarah is currently director of SPACER – a purpose-built space for transdisciplinary practice. She is an Honorary Research Fellow at the Centre for the History of the Sciences, University of Kent, and a visiting lecturer at Central Saint Martins. Sarah is co-curator of the Science and Arts Section for the British Science Festival. More on her work can be found at http://www.sarahcraske.co.uk/

Charlotte Sleigh is Professor of Science and Humanities at the University of Kent, UK. Originally a historian of science, her work now touches on science in its relation to literature, art, and communication, amongst others. It is her understanding of science as a social process, mediatised through cultural forms, that underlies these interests. She is the author of numerous books and editor of the *British Journal for the History of Science*.

References

(1) Sleigh C and Craske S. Art and science in the UK: a brief history and critical reflection. *Interdisciplinary Science Reviews*. 2017;42: 313–330. Available from: doi.org/10.1080/03080188.2017.1381223.

(2) Sleigh C, Craske S, and Park S. Lowering the tone in Art and Science collaboration: An analysis from Science and Technology Studies. *Journal of Science & Popular Culture*. 2019;2: 37–51. Available from: doi.org/10.1386/jspc.2.1.37_1.

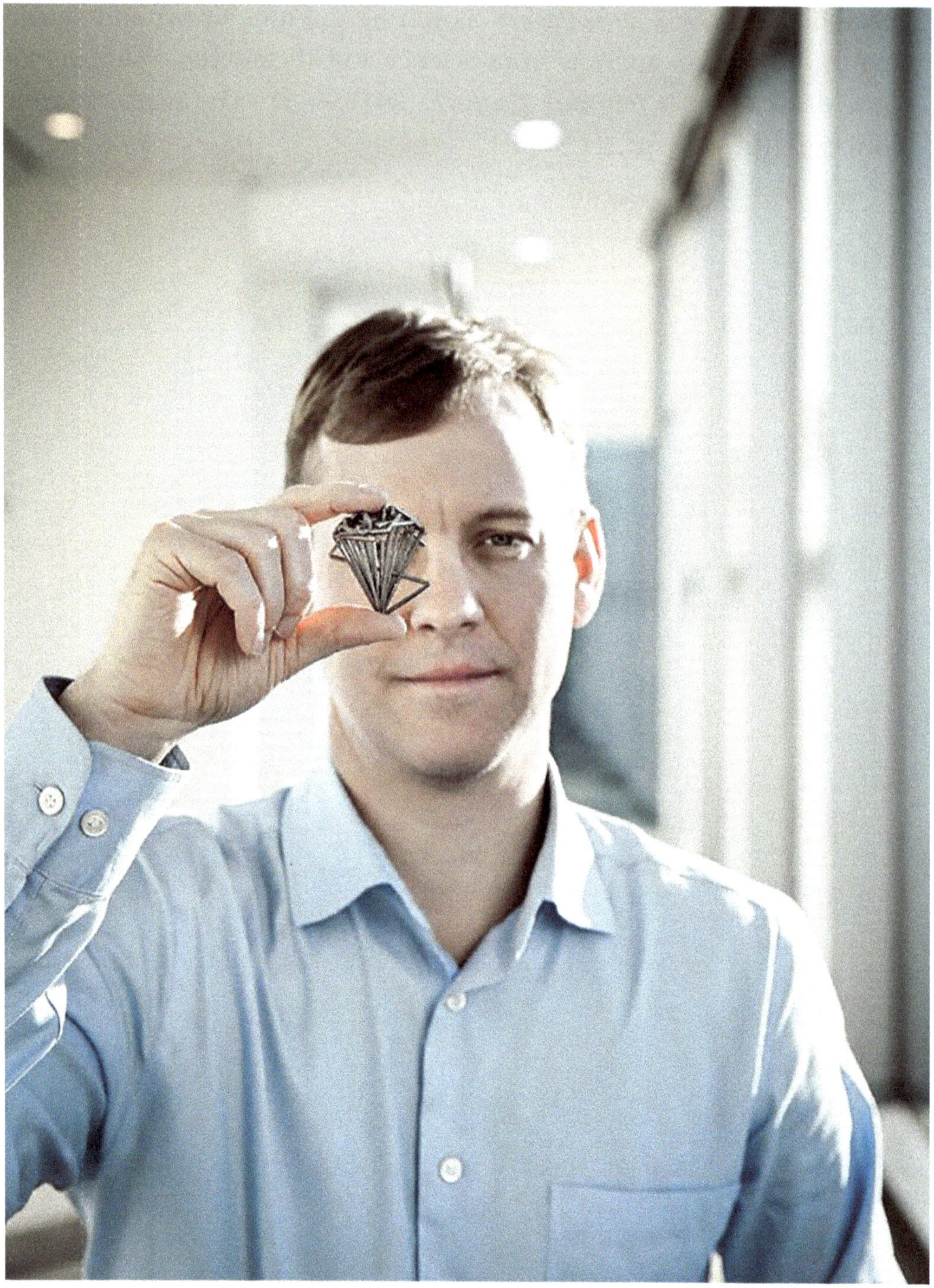

Wayne de Fremery, `Reprinting ``Azaleas'' (2015).

Literary Phenomena and Alternative Encounters

Wayne de Fremery

When we think of books as mere objects, we lose sight of them completely. If we treat texts as woven systems of physical, human, and social agency, their unending generative nature comes once more to light. In this interview, Wayne de Fremery explores what it means to think about texts and textual transmission as living processes.

Your research concerns bibliography and the socialisation of 20th-century Korean literary texts – a line of questioning that challenges traditional notions of texts and textuality. What does it mean to conduct cross-disciplinary research into Korean Poetry?

My work concerns the poetics of documenting literary phenomena and the ways in which the 'literary' can be investigated as lived experience. This means I work across a range of traditional literary and bibliographic scholarship, but also that I conduct artistic experiments aimed at creating new methods for documenting the elaborate technological and cultural systems that iterate texts (with a particular focus on Korean poetry). My aim is to ensure that the texts of Korea's oral and print traditions remain alive and recognisable in the lived experience of those who grew up in eras when literature was synonymous with certain technologies, such as manuscript and print technologies, while also inspiring those who will need to keep Korea's texts alive as meaningful expression in media yet to be imagined.

My doctoral dissertation concerned documenting books of vernacular Korean verse produced initially in the 1920s – a time of great political tumult on the Korean peninsula and generative poetic experimentation. My aim was to describe what a well-known bibliographer, D.F. McKenzie, called the sociology of texts. I tried to describe the linguistic content of the books and how they worked as literary phenomena, which necessitated describing the people who made them – the poets but also the publishers and printer and distributors – as well as the technologies they used – the kinds of presses, the variety of typefaces, and the machines used to cast the type. The simple argument that I attempted to articulate was that literary analysis is premised on assumptions about the material iteration of literary texts, many of which, in the case of Korean poetic texts from the 1920s, were incorrect because scholars had not carefully investigated how books of Korean poetry were created. Mandated by my university, I used print media to document my engagement with roughly 40 books of vernacular Korean verse and roughly the same number of periodical issues that contained the poems of a poet I am particularly interested in.

I am freer these days to experiment with the media I use as documentary tools for my engagement with Korea's texts; my more recent work has focused on documenting the social and technological systems that iterate Korea's textual record in computational environments. For example, I have experimented with collaborators on a variety of methods for describing the coding standards that underpin the expression of Korean texts in digital environments. I have also attempted to creatively document the poetic structures of Korean poems by mapping them onto visual structures and colours, including in three-dimensional virtual reality (VR) space – an investigation into alternative modes of documenting literary phenomena and expanding the palate of technological tools for bibliographic expression. These projects reflect my interest in how alternative forms of engagement with literary phenomena might enliven or detract from people's lived experience with literature: Would these alternative bibliographic expressions prompt wonder? Would they inspire and delight? Would they instruct, and if so, how?

A figure you have invoked in your work is Donna Haraway's `cyborg' – a stand taken against the Enlightenment's separation between Subjects and Objects, and one that explores a more intimate understanding of how conceptual, bodily, material, social, and digital forms interact. Can you tell us more about this figure and how it frames a reappraisal of poetry and the producers of poetry?

Michelle R. Warren points out that 'analogies have often been drawn between the human body ' and 'physical text[s]' (traditional Subjects and Objects, respectively). 'The metaphor shifts significantly ', she writes, 'with the image of the 'cyborg' body posited by Donna Haraway…. The cyborg challenges naturalised genealogies of (textual) transmission from generation to generation, underscoring the body's construction through purposeful interventions ' (1: p.130). Let us take this apart in steps.

Bibliography concerns the 'writing out' of 'books' (from generation to generation). It has traditionally been understood as the scribal practice of copying and then as the study of how books came to be 'written out', i.e., all the technologies and social practices associated with producing texts. Underpinning these practices was the belief that the texts, the human bodies, and the technologies producing them all were alienated from each other. Haraway's cyborg is an opportunity to productively complicate these separations between Objects and Subjects that have been guiding bibliographic practices. If a 'text' is not separate from its material shapes or its conceptual forms, and the experiencing body is technologically hybrid and not separate from the technologies of textuality, then we can think about text and textual transmission differently. Recognising their interrelation, as Haraway's cyborg helps us to do, reorients how we might think about 'the copying out' of 'books'. We can think about the ways in which texts and human bodies are both, to varying degrees, made through 'purposeful interventions'. We can think from an alternative vantage point about how we might intervene in the technological production of texts and the naturalised bodily practices associated with textual experience and then experiment with new methods for 'copying out' 'books'.

Karen Barad has a useful term: 'intra-action' – interactions that do not assume *a priori* relationships between documentary apparatuses and the phenomena that they document. Thinking about textuality as the intra-action of human bodies and textual technologies expands the possibilities for imagining literary and bibliographic expression. We can understand texts as woven socio-biological/ technical institutions that prompt and enable expressive cultural memory. And we can rearticulate common metaphors for describing textual experiences, such as getting 'lost' in a story or being 'immersed' in a book, by making the metaphors literal experience through designing experiences that can be inhabited by the body in a theatre space or a virtual/augmented reality environment. Creating metaphors that can be inhabited in this way helps emphasise the ways in which the body then can be understood as a medium through which a text can be copied out. We do not commonly think of reading and bibliographic practice in these terms. Juxtaposing these new bibliographic environments and modes of expression with what has come before helps to emphasise textual production as a bodily intra-action with language as it is materially expressed.

Theorising texts in this manner also helps to expand and productively reorient thinking about specific textual formations we have typically associated with particular places or groups of people, as is so frequently done with the study of national literature. If we think of texts as woven systems created by biological/socio-material intra-actions, then we are able to investigate them without falling into confining and frequently essentialist definitions of peoples and cultures. We can investigate the marvellous specificity of texts woven at particular historical moments by people interacting with any variety of material or conceptual technologies. We can also imagine alternative intra-active experiences and build a means for exploring the ways in which texts can be experienced in the future. For example, when ‘Korea’ – understood in Leigh Star’s terms as a boundary object – is expanded beyond the strictures of the nation, texts associated with it become more interesting and more vital because we can think more creatively about what might be included within the boundaries of ‘Korea’. 20th-century definitions of Korean literature as documents composed by people of certain national/ethnic categories using specific technologies, such as the paper page and han’gu⊠l orthographic systems, can be refigured so that what is thought of as ‘Korean’ can be inhabited by many more people using a wider variety of technologies of knowing. This is hopeful because it means that there can be many more ways to inhabit and reiterate ‘Korean’ literature now and in the future.

Is the concept of `becoming media' suggested in cyborg textuality – a vital phenomenon in which work, author, reader, curator (and so on) are understood to participate in an ongoing process of becoming something new together?

Sure. The idea of ‘cyborg textuality’ and ‘becoming media’ can be related – both are productive metaphors for investigating the woven nature of media. They are potent because, as your previous question suggests, they enable us to investigate experience with what we call ‘texts’ and ‘media’ with a schema that places emphasis on material details and conceptual formations that we may not have previously noticed or considered. Conceptual frames always tune us to notice certain elements of experience and not others. With each investigatory metaphor, we gain a new tool as well as new kinds of evidence. There are two important points to remember, however: The first is that the investigatory metaphor will only allow certain elements to be noticed; the second is that the metaphor does not need to be based on any ontological belief, philosophical outlook, or conceptual system. Bibliographical study has often taken science, and particularly biological science, as a guiding conceit. (Consider stemmatics: The practice of discerning the development of a textual work by tracing the relationship between different textual witnesses, an approach that derives its analytical power from Darwinian evolutionary theory.) Noticing this is important because it allows us to frame the idea historically and consider if we would like to consider using the same metaphor as an investigatory tool. We are reminded that we can make our own conceits. Whether it is humanistic ideologies about human experience or post-human beliefs about hybridity, our philosophical engagements can be fantastically powerful tools for revealing facts about our experience. But, I would stress, we can creatively make up our own conceits and use them as the conceptual infrastructure for organising our investigations and revealing additional facts we might puzzle over in wonderment or stand beside in awe.

With a focus on innovation in material and communicative forms, your work explores the facets of a textual source as physical (an inscription in a book), logical (as something on which operations can be made), and conceptual (through which a work can be understood). What is the origin of this approach, and how are you developing new relationships between these different facets?

The tripartite notion of an object having three inheritances – a physical inheritance, a logical inheritance, and a conceptual inheritance – was initially articulated by Kenneth Thibodeau to describe digital objects (the physical processes of electronic inscription that make up the storage systems of computational systems). I find Thibodeau's conception useful because it enables us to see digital documents as a variety of processes interacting. The physical components and processes of digital objects are inherited by logical systems that are, in turn, acted upon to create a conceptual object. The conceptual object in Thibodeau's schema is what we would see on a computer screen or projected onto a wall. Seen as interacting processes that are simultaneously conceptual, logical, and physical, digital objects become wonderful tools for investigating the 'logic' of our 'conceptions' and their relationship to physical structures that are often obscured from view.

A variety of encoding systems have been developed to allow computational systems to display written languages consistently. These systems are, in part, what software works on when we 'word process', for example. Although there is nothing except the powerful logic of our historical experience to suggest it, text is displayed in these systems using traditional orthographic systems as conceptual objects that look like a printed page. Understanding the process of materialising texts as conceptual objects on our computer screens allows us to play around with their logic to create alternative conceptual objects. These alternative conceptual objects can be used to illuminate aspects of our textual experience that are less visible but, nonetheless, crucial to how we interact with a text. They can also teach us things about our texts that we could not have known otherwise – simply because we had not yet created a conceptual shape for the new discovery.

We can transform the conception of a poem, for example, as something printed on a page into something that can also be iterated in the shape of a tree or as a pendant hung from our neck. This, in turn, enables us to ask provocative questions about poetry and literature, such as whether a poem always needs to be iterated by the conceptual models of print or orality, as they have been traditionally conceived? We are presented with the productive challenge of finding a conceptual shape for poetry that can honour its long association with print technologies while simultaneously orchestrating new poetic experiences by means of alternative technologies.

My work to create poetic experiences that can be inhabited as theatrical experiences modelled on the bibliographic and linguistic cues associated with Korean poetry is one example of this approach.

In creating innovative digital objects from poetry (such as navigable virtual environments), you are introducing conceptual models that may also obscure or negate the original in some way. In what spirit would you wish the user to engage with these new works?

I would hope that these new conceptual models are opportunities for wonder, inspiration, and contemplation. My goal is not to obscure or negate previous iterations or anything that might be considered 'original'. Rather, I attempt to use the power of defamiliarisation to heighten a person's understanding of the expectations they bring to an encounter with an object. The goal of fashioning new digital objects from poetic texts is to enable those encountering these new objects a clearer understanding of the expectations they brought to the process of reading poetry and the individual texts that they expected to read.

For example, with collaborators, I have designed and built immersive theatre and VR experiences that express books of Korean poetry as navigable forests, mapping the stanzas and lines of individual poems to different tree structures. The book of poems we modelled was initially produced during the Japanese colonial occupation of Korea (1910–1945). Poems from it are frequently taught in Korean middle and high schools. As a consequence, readers approaching the poems have many expectations about the book based on their educational experiences and general beliefs about Korea's colonial experience. Engaging the book expressed as a forest necessarily thwarts these expectations and allows participants the opportunity to reassess their beliefs about the book and perhaps, more generally, what 'reading poetry' might mean.

It is important to note that the theatrical and VR experiences are explicitly designed to allow those who inhabit them to navigate through the forest 'toward' more familiar, and earlier, representations of the book. A user wishing to see the digital text of a poem used to model a tree in the forest can touch a tree in the environment and see the digital text. If a participant wishes to see images of the printed text upon which the digital text was based, s/he can navigate through the digital text to images of the book's first printings. There is no way to reach the ' original' – whatever that might mean – in the environment. It is hoped that the journey toward the idea of an origin inspires wonder about the book as it was produced initially, and curiosity in how the book has been iterated since its initial production (by individual editorial, social, educational, and any number of technological systems), so that we can think creatively about conceptions of historical periods, what constitutes poetic experience, and how we might reiterate the book into the future.

If these new relationships being created between physical/digital objects and environments are, in essence, exploratory and not yet complete, where do you see the potential for new conceptual models to emerge?

New conceptual models will emerge through detailed and creative exploration of the ways in which we create and interact with the socio-technological infrastructure of what we call texts, since texts now are at once physical/digital objects/environments. Our textual systems are, and have been, so complex that we only capture a small portion of their complexity with any textual presentation or description of a textual event. The work of making new conceptual models will entail getting dirty with the material details of the systems and procedures that create textual experiences in order to have better bibliographic and artistic control over the breadth of textual complexity and its expressive power. I should stress that

I do not mean anything like monopolistic control over textual production when I talk about bibliographic and artistic control. Rather, I mean the ability to see the complexity of textual expression with greater precision and understanding so that we can better choose the creative constraints within which we choose to create and document textual experience. New conceptual models will be generated by those neck-deep in the material and conceptual minutia of textual experience groping for a way to describe and express the rich complexity they discern.

These new models have a built-in capacity to support probing questions into Korea's cultural record and enable direct action of different kinds in turn. Can you envisage new forms of memorialisation, political action, and cultural discourse emerging from your work?

Yes! The ways in which Korea's cultural record is curated will determine how Korea's past is memorialised, what kind of political action becomes tenable, and what shape cultural discourse will take going forward. Any method we use for organising, preserving, and reiterating materials associated with Korea's cultural past needs, also, to be understood as a political action. The ways in which we can imagine the unfolding of these process are, of course, manifold, although it is difficult to convince people of this truth ; I suspect this is because we think of curating the cultural archive as a kind of bureaucratic process associated with paper-based practices, rather than a lived, artistic engagement that can be conducted in any media.

My work aims to focus attention on these facts so that there can be a healthy, informed, and creative debate about ways we might sustain Korea's cultural record in the present – so that memory practices, cultural discourse, and political action can serve truth and the health of individuals and communities. An individual's sense of well-being is tied to her or his sense of belonging – to a partner, to a family, to a community, to a place, to a tradition. And there is no greater danger to a person's well-being than the revelation that one's sense of belonging has falsehood as its foundation. Those of us working hard to curate this record face some significant challenges (as well as some fantastic opportunities). A variety of forces threaten materials created during Korea's colonial occupation, for example. Korea's literature was frequently printed on highly acidic paper, and that paper is now burning itself up. Ironically, this might not be the largest problem in bringing this cultural record alive for people; neglect – whether wilful or tacit – takes that prime position (for example, when linked with lingering resentment over the colonial occupation).

The huge amounts of digital information created since the mid-1990s present another enormous challenge to curators of Korea's cultural archive. As far as I know, there is no archive of historical software anywhere in Korea that will enable curators and archivists, let alone the public more generally, to access digital documents created even five or ten years ago, let alone twenty or thirty. Nor are there truly long-term plans for sustaining access to the rapidly expanding datascape that constitutes the cultural interaction and exchange of our present historical moment. If not addressed immediately, there is a good chance that, fifty years from now, we will be looking back at a Dark Ages caused, ironically, by the digital brilliance of our contemporary experience and our short-sightedness.

At an institutional level, you are creating new fields of research. What are they, how have they come about, and what is their relationship to more established disciplinary traditions?

Computational bibliography and the sociology of data are phrases of my invention, and the working title of a book I am writing. 'Computational Bibliography and the Sociology of Data' proposes to expand the scope of what bibliography describes and to diversify the forms used in bibliographic description. As I have been describing, and as an etymology of the word ‘bibliography’ suggests, bibliographers in the past used bibliographic forms – books – to document and investigate books. ‘Computational Bibliography and the Sociology of Data’ suggests documenting computational systems using computational systems. Since computational systems and the data they express are social entities, this documentary practice is a kind of sociology – one that is facilitated by new forms of descriptive bibliographic expression that aim to pre-figure and constitute fruitful methods of scholarly investigation and meaning making. How, for example, might we use the creative force of computational systems to document the proprietary software systems – including all the people and technologies – that iterate the books we read on our electronic devices or the algorithms that suggest books we might like to read? As my book argues, answering these questions is a matter of attending to and documenting the technologies and the communities that use them. Those familiar with bibliographic research will recognise that my book is deeply indebted to D.F. McKenzie and his ‘Bibliography and the Sociology of Texts’, which helped to reorient and expand the focus of bibliography in ways that are still productive. This is especially true now when what we mean by ‘book’ is evolving and computational systems as expressive mediums present us with new opportunities for documenting our changing relationship to what we call ‘books’.

This bringing-together of different methodologies has led you to combine forms of geospatial and textual analysis to bear on the different forces at play in shaping the production of texts. Can you tell us more about this research and your findings?

‘What we call literature is an institutional system of cultural memory ’, writes Jerome McGann (2: p. ix). Since I am interested in what our systems of cultural memory evoke and *how* institutions of cultural memory have been shaped, there is no avoiding the role played by place in literary formations:

Memory and its institutions are always spatial – practically and imaginatively. What we call literature helps us to imagine or reimagine the contours of the most familiar street of our hometown, or what might constitute a 'street' in alien civilizations. Literary studies, of course, frequently organises texts according to ideas about national boundaries and characteristics – 'Eastern' literature, 'American' poetry, 'South Korean' fiction, and so on. These generic categorisations can generate meaningful ways to know texts. My work attempts to provide additional ways to know textual bodies, such as through mapping the people and materials used to create our institutions of cultural memory with a kind of forensic precision that has not been previously attempted.

While working on my doctorate, for example, this meant collecting and organising information about the places where books of vernacular poetry were physically produced. Japanese law required the names and street addresses of publishers, printers, and distributor(s) to be included in the colophons of all books. This was a security measure that enabled Japanese police to find anyone printing subversive materials. Publications would also include information for book buyers, such as bank account details and contact numbers. By organising this information, I discovered something previously thought to be outside the realm of literary study: I learned that more than half of the books of vernacular poetry produced in the 1920s recorded by Korean bibliographers were created at one geographic location, with more than a third produced by one man, No Ki-jo⊠ng, working at printer and publisher Hanso⊠ng Toso⊠ Chusikhoesa at Kyo⊠ng-so⊠ngbu Kyo⊠nji-dong 32-po⊠nji (in what is now called Insa-dong in Seoul). Here we see how one individual can play a tremendous role in shaping cultural memory through the ways in which poets and other literary artists express themselves with the technologies available at a given historical moment. This makes clear that concepts such as 'Korea' and 'Korean literature' are best shaped in productive tension with such specificities in mind.

My more recent work attempts to map the people, places, and institutions that are iterating Korean texts in computational environments. I am curious to know if there is a place comparable to Hanso⊠ng Toso⊠ with someone like No Ki-jo⊠ng overseeing its production. Initial indications are that there is not. While much of Korea's printed literature is made in places such as Paju, a city near Seoul where many of the important publishers and printers have recently relocated, my sense is that the geography of 'Korean' textual production is quite global. Korean authors who split their time between Seoul and New York will use software adhering to common international standards to create manuscripts for publishing houses in Paju, New York, and London (with portions posted online using social media platforms operating out of Silicon Valley and Pangyo). The challenge is finding ways to document and map these geographies so that we can better understand the ways that we know ourselves in our history and the places we inhabit – a challenge that necessitates new kinds of cartographic tools.

With the development of innovative approaches comes an active negotiation with disciplinary, cultural, and social norms. Do you see these new models as capable of driving innovation in scholarship? As such, how have others responded (both within Korea and beyond) to these new ways of working?

Will the models and methods I am investigating drive innovation in scholarship? I hope so! I would also hope that they drive innovation in the arts and in the ways we craft our institutions of cultural memory. But, as the framing of your question suggests, any such effects will require long-term disciplinary, cultural, and social negotiations, a process that will also take from me the methods and models I am developing – if indeed they were ever 'mine'. This process was hard for me to come to grips with at first. I used to be annoyed when those in a given field (whether literary studies, bibliography, design, or information science) became uninterested when I talked about practices outside of their immediate interests. Poets and artists thought of me as a scholar because I would ramble on about historical and theoretic minutia rather than devoting myself to *making* art. Scholars thought of me as a quixotic poet and artist who spent too much time designing books and *playing* with technology. It was only later that I understood that my job, however defined, was not just to persuade people of the value of the ideas I presented but to enable others to adopt the ideas as their own. I realise now that I was not giving researchers in information science, for example, Korean poetry as something that could enliven their practice; nor was I enabling those in literary studies to take bibliography and various practices in media studies to enliven the study of literature.

There are small indications that my ideas are becoming less and less my own, which makes me hopeful. Scholars in information science, for example, are increasingly interested in data and perspectives from literary and cultural studies. Those in literary and cultural studies recognise that computational systems are profoundly changing the ways in which we form our cultural experiences and memory. Bibliography and literary studies, what were once called 'lower' and 'higher' criticism, are increasingly approached as an integrated field by scholars. Moreover, I sense that evolving ideas about the art of scholarship and the scholarship of art are reshaping institutional practice and the ways that scholars and artists express themselves. Traditional formations such as the journal article and the academic monograph are now being interrogated, with many beginning to wonder if they are the only standard by which academic achievement can be measured. Those of us who love these forms are excited by this development because we can continue to use familiar expressive practices while also imagining new ways to express our discoveries.

Wonderfully difficult and vitally important questions about our beliefs concerning the boundary between scholarship and artistry are posed when artist-scholars culture human cells for aesthetic ends and engineer RNA sequences to produce proteins encoded to spell out poems. I sense that some scholars would not take offence if they were called artists. Similarly, I think fewer poets and artists would be offended if they were called scholars and scientists. In short, I am excited by the possibilities these hybrid orientations present while cognizant of the fact that boundary making and the evolving practices that have individuated scholarly from artistic practice, as well as academic disciplines from each other and 'industry', have often served immediately useful and vital purposes. It is an exciting time to be curious.

References

(1) Warren MR. The Politics of Textual Scholarship. In: Fraistat N and Flanders J. (eds.) *The Cambridge Companion to Textual Scholarship.* Cambridge, Cambridge University Press; 2013. p.119–134.
(2) McGann J. A New Republic of Letters: Memory and Scholarship in the Age of Digital Reproduction. *Cambridge,* Mass.: Harvard University Press; 2014.

Author Biography

Wayne De Fremery is an associate professor at Sogang University in Seoul, in the School of Media, Arts, and Science. His research concerns bibliography and the socialisation of 20th-century Korean literary texts. Wayne holds a bachelor's degree in Economics from Whitman College, a master's degree in Korean Studies from Seoul National University, and a doctorate in East Asian Languages and Civilizations from Harvard University. Books and journals designed by Wayne have appeared with the Harvard Korea Institute, the University of Washington Press, and his own award-winning press Tamal Vista Publications. He holds three patents for inventions that creatively document and preserve culture and, in 2017, became a member of the Korean Agency for Technical Standards – the first non-Korean citizen to represent South Korea on an ISO committee. More on his work can be found at http://www.pwdef.info/index.html.

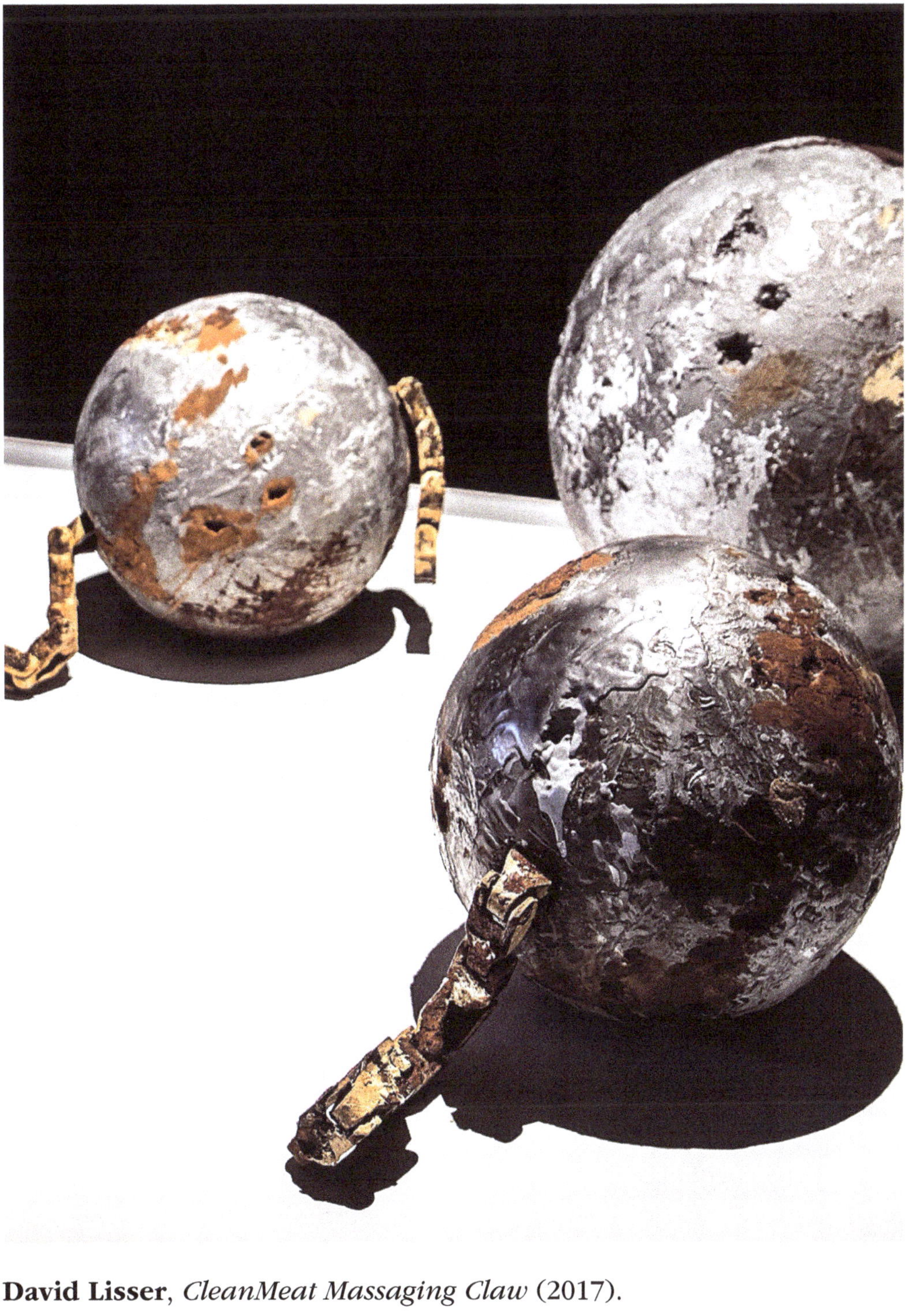

David Lisser, *CleanMeat Massaging Claw* (2017).

All Bets Are Off

David Lisser

Global food security is becoming a pressing issue of our times; we need to respond, but all bets are off concerning the actions we should take. In this interview, David Lisser explores through the `The CleanMeat Revolution' (1) what a `future historical retrospective' might reveal about the path we take now and the unexpected events that may befall us.

It is the year 2120, and in your current historical research you are exploring the rise and fall of *in vitro* (cultured) meat products around the middle of the last century. What were the challenges around food security that first emerged at the beginning of that century, and how did they prompt a flourishing of different food solutions prior to the dominance of the `CleanMeat' movement?

At the beginning of the 21st century, popular opinion held that the predominant global challenge facing food production was a warming climate and the inconsistent growing conditions that this caused. Although this is accepted as a primary driver, many other interacting factors can be identified, so producing a more complex and nuanced picture. There are too many individual causes to go into detail here, but academics now accept that a cocktail of factors contributed to the emergence of worldwide food insecurity, including: rising global temperatures (with effects on cycles of drought and flooding), social upheaval related to changing patterns of migration and displacement, systemic failures in the form of resource mismanagement, the production of excessive waste, the over-reliance on biofuels, and cultural factors such as the status-driven increase in the consumption of meat-based food products.

The common Malthusian pre-conception that we simply could not feed a growing population has proven incorrect. For decades, we have produced far more food than the world's population could actually consume, but, due to critical imbalances in the distribution of power within the food system, and a lack of regulatory oversight to tackle such market failures, around a third of food production has been lost to waste, with the remaining two-thirds unequally, and even unhealthily, distributed. By the turn of the 2020s, the inequalities built into this system were becoming more readily apparent: Nearly a billion people malnourished in a world sustaining over two billion classified as overweight or obese. This points us towards the rise of CleanMeat and its immediate precursors – all attempts to stabilise global food production and consumption.

A note on taste: The watchword for the diet of the 2010's was protein. In the west, the ideal body image, for both sexes, had shifted from skinny to muscular; in developing economies, rising incomes resulted in a greater demand for meat products. The rather lazy characterisation that the 'protein-obsessed' people of the West demanded high-welfare meat and/or vegan alternatives, whilst those in China, India, and other growing economies were unfussy about sourcing policy, was popular at the time, but proved untrue on closer analysis. The breakdown on meat consumption globally reveals that industrially prepared meat was popular in early 21st century regardless of the country in question, and that cost was a primary determining factor in the choice of meat product.

With the emergence of CleanMeat, we saw the creation of a new living medium through innovative forms of techno-cultural intervention – a non-reproducing form of life dependent on human activity. What was the `living' status of clean meat in this form (a form removed from any known natural ecology)? Further, what was the consequence of CleanMeat's development for traditional Livestock of that era once the pressure of human domestication and selective breeding was removed?

The 'Livingness' of CleanMeat was initially a controversial subject, and one that large commercial producers spent a great deal of time and money negotiating. A series of creative public awareness campaigns were successful in persuading the public that CleanMeat was essentially a 'natural' non-animal product in its own right. Their efforts were helped by the fact that, since the early 2010 s, PETA (People for the Ethical Treatment of Animals) had been funding research in this area and were vocal advocates for the positive, ethical implications of commercially available CleanMeat. One of the more successful campaigns in the West showed a tour of the CleanMeat production facilities (termed 'carneries'), comparing the manufacturing process to that of brewing beer ; this played heavily on a comparison between the living cells used in Clean Meat (originally derived from animals) and the status of yeast in the brewing process. Fermented drinks have been produced for millennia, so this comparison helped CleanMeat attain a high degree of cultural normalcy. In alcohol-abstaining cultures, the campaign was tailored to draw comparisons with yoghurt production and other fermented products. Any allusion to foetal bovine serum was carefully avoided.

All the while, the emergence of full-scale CleanMeat production did not spell the end for traditionally reared livestock. Throughout the 21st century, farmers continued animal husbandry practices along the lines of previous generations. In fact, because CleanMeat took such a large market share from industrially farmed animals, there was a resurgence in low-yield and high-welfare livestock practices (Slow Livestock). Free-range and organic became the new standards for 'real' animals. A two-tier market emerged, whereby CleanMeat filled the requirement for cheap, healthy meat, whilst traditional meat stocks acquired the status of a luxury product. Arguably, although CleanMeat significantly reduced the number of animals raised at any given time, it did drive considerable improvements in animal welfare for those artisanal farming operations that remained. Indeed, this shift from large-scale to specialist production drove an overall change in the labour market – with interesting consequences. As industrial farming had become so heavily automated by the late 2020 s, the scope for further redundancies in that sector was fairly limited. So, although a handful of major producers did lose business, a significant number of farmers intensified their focus on traditional techniques and the luxury meat market, with the effect that the number of skilled workers in animal husbandry, butchery, and meat preparation actually increased over this period.

Looking back at the predictions of the 21st century around food security, was the meteoric rise of CleanMeat and its equally rapid collapse as a food source in anyway predictable? Can we learn anything about the course of disruptive innovation?

CleanMeat promised a form of meat production that was less water-, land-, and energy-intensive and resulted in negligible GreenHouse Gas emissions. This was highly desirable for major food producers, with the wider potential economic and environmental benefits being a key target for government policy. That it all but eliminated animal suffering within its own supply chain was not in itself valuable to the industry, but was considered, none the less, a highly marketable concept. During its early development in the 2020s, a number of nation states identified key values in the idea of self-reliance in meat production. Short supply chains and increased control over the whole production process reduced a reliance on global markets and food aid, enhanced food security, provided greater control over food safety regulations, and reduced the threat of food terrorism. As a politically powerful message at the time, this helped bolster a rising nationalism across the globe. Enormous sums of money were invested in the race to develop commercially viable and palatable cultured meats. As the cost of production tumbled, research activities began to focus on overcoming long-standing issues around food texture and the wider public perception of this new, innovative product.

Food fashions are inextricably linked with socio-economic status, and, initially, it was the middle-class that took to CleanMeat most enthusiastically, lured both by its environmental credentials and rejection of animal suffering. It soon became an aspirational product, driving product innovation that targeted a broader range of socio-economic groups. The ease with which CleanMeat came to dominate the global market was truly unprecedented, but also a contributing factor to its eventual downfall. During its emergence, CleanMeat was one of the most heavily monitored industries in the food sector – arguably, a necessity in gaining consumer trust. But, as the demand for sector growth and cost-reduction increased, a period of deregulation ensued (secured by lobbyists and environmental groups), followed by market failure.

For instance, cases of antibiotic resistance had occurred only infrequently in its initial period of development; quickly identified, these high-risk products were prevented from coming to market. As the regulatory environment was relaxed, however, incidences of resistance began to increase and a small number of products carrying resistant bacteria entered into the food supply chain. Only a few deaths resulted, but this was to have an enormous impact on public trust. Mainstream and social media heavily publicised the deaths, with investigative journalists soon uncovering mixed donor cell-reactors; although as safe as many other widely consumed products, the so-called pig-cken meat was considered unacceptable. The smaller clean-fish, mocktopus, and crustacea industries fared worse – small quantities of cockroach-derived stem cells were found in prawn-sticks and lobster meat. There was public outcry.

Some historians have compared this wave of public mistrust to the BSE crisis of the late 20th century in Europe. A more apt analogy, perhaps, would be that of the banking crises that occurred in 2008 –2010, 2036 –2040, and, most recently, the global freeze of 1989. We can readily discern the perennial favourites of the economic historian: market failure driven by deregulation, irregular competition practices , misguided environmental economics, the collapse of a shared-resource systems (the tragedy of the commons), and so on.

CleanMeat was eventually replaced with the fully synthetic protein substitute Synthein. How was it an advancement on its predecessor in terms of customer appeal, customisation, and business model? As such, is Synthein the perfect technological solution, or do you see a more fraught pathway ahead for this new product?

Although Synthein largely replaced CleanMeat as the primary source of cheap protein, it must be understood as a very different kind of product, with a very different sourcing chain. Synthein producers supply the base growth medium, but then, for the majority of consumers, product maturation, flavouring, and harvesting are done in the home or on a slightly larger scale at community co-ops. The development of sophisticated flavour coding modules suitable for home use proved an absolute game-changer for the industry. Even at the peak of CleanMeat's market dominance, production techniques were unable to recreate the subtlety of natural meat's flavour and aroma; indeed, due to inadequacies in the waste removal systems, many of the cheaper, unrefined CleanMeat forms had a faint, but pervasive, tang of urea. It would be very hard for us to accept that today, but, at the time, many considered this an acceptable pay-off for such a cheap protein source. And, besides, all but the most prohibitively expensive natural animal meats had, by that time, been engineered towards a rather homogeneous flavour palate.

Synthein producers initially worked alongside Michelin-star chefs, social innovators, and haute-vlogueurs to develop celebrity-endorsed flavours. Once the domestic flavour-synthesisers became available, however, many more became self-styled TJs (Taste Jockeys), 'laying down' new tastes and olfactory experiences. Although the application of these new flavours to CleanMeat was attempted, it proved near-impossible to chemically bonded flavour compounds to the product. Whilst it is still not fully understood why, in the few successful cases that were documented, the newly applied flavours were reported only to exaggerate the product's underlying urea-characteristic.

In one sense, Synthein producers were successful in developing a business model adopting the best of patent protection whilst embracing an open-source movement that could secure its status as an endlessly customisable and bespoke product. The core processes behind substrate growth and cell proliferation remain a tightly guarded secret, whilst customisation activities are not only accepted but also actively encouraged. The many competitions annually to celebrate flavour innovation are hotly contested affairs. There are risks, however, associated with the Synthein approach. It is theoretically possible to hack the cloud-connected taste-synthesisers ; so it may not be long before system vulnerabilities are identified and exploited (whether for a prank or to more malevolent ends), weakening the brand's current dominance. It seems almost inevitable that this sector will suffer from the same pains originally experienced in the smart home industry.

With the development of this fully vegan – indeed non-life based – food source (Synthein), have we seen public attitudes changing towards other life-forms and the ethical relationships we form with them in our shared ecosystems?

We have seen a gradual realignment in our relationships with nature; although the development of synthetic food has played a key part in this, it is the unprecedented consequences of climate change that has proven a primary driver in the modification of human behaviour towards the environment. Early efforts to combat climate change were, of course, too little and too late for Osaka, Shanghai, and Miami, but the devastation of these floods did mark a gear change in the seriousness of international response to our changing climate.

When I tell my students about the conception of *natural resources* in the mid- 20th century, they can hardly believe their ears! The notion that our fragile ecosystem was understood in merely financial terms is as alien to them as our current relinquishment targets would be to national politicians of the late 20th century. Indeed, we have come a long way: Concepts such as deep adaptation, sustainable intensification, and whole-world health – once popular only amongst the educated, wealthy, and liberal as value-signalling conversation pieces – are now much more pervasive and commonly held. Our de-growth strategies do appear to be helping a number of key natural habitats begin the long road to recovery, and, yes, a good proportion of global food consumption is now completely synthetic. These are considerable achievements and should be lauded.

However, as our relationship with the environment becomes increasingly the subject of global, algorithmically driven auto-responses, inherent problems in the system may emerge. Although designed to moderate political and financial interests in global decision-making, the very architecture of these decision systems has come under scrutiny. Some argue that the relationships between data collection systems and auto-response outputs reveal a misplaced philosophy of control and management of our biosphere. I am a proponent of updating these systems to better incorporate nuanced and adaptive decision protocols that emphasise our forms of co-existence within the world. Whether there is the will to see through such changes remains an open question.

Finally, is it fair to say that artists working with living media at the turn of the 21st century in some way played a role in the development of these new, secure food sources? To what degree did they inspire advances in the field by showing what was possible, whilst also offering dissenting and ethically motivated accounts of the dangers involved?

In the famous Greek myth, Daedalus crafts wings for himself and his son and warns Icarus not to cruise too low so that sea-spray will dampen the feathers but also not to fly too high lest the sun melt the wax that binds the wings: Icarus should keep a steady path, striving to avoid both indolence and arrogance. It could be suggested that, in the story of CleanMeat, early practitioners working with living media acted the part of Daedalus. Even the very name, which translates roughly as 'cunningly wrought', suggests a manner of working that navigates the possibilities of nature to bring forth new realities. Daedalus created his wings in order to escape from the legacy of his previous invention – The Labyrinth; our practitioners created art in order to help others escape from an overbearing legacy of the scientific method. The new forms that emerged from these practitioners were, therefore, points of contestation by their very nature: They were not created simply to showcase new methodologies with pre-ordained aims clearly in sight, but rather as working prototypes that, through their mere existence, interrogate the social and ethical potentials of research.

Early practitioners showed what was possible and, in the manner and context of presentation, challenged future adopters to carefully consider the consequences of pursuing such possibilities. Business leaders and entrepreneurs took enthusiastically, by and large, to these new models of working but less so to the challenging conceptual implications that came with them. Hindsight is a wonderful thing, and if we could show those early designers the fruits of their labour I am sure we would witness both elation and horror. For example, the development of semi-organic, self-growing buildings marked an important shift in the construction industry from a damaging reliance on mineral extraction, but, as the recent documentary '*Increville – the City that ate itself*' revealed, there were a number of unforeseen complications with such large scale bio-construction. Similarly, the new food sources originally conceived in the late 20th century came under the banner of contestable design – artists and experimental designers working at the very edges of disciplinary practice to imagine extraordinary responses to complex, contemporary challenges. Some years later, and after adoption and reform by Industry, this work can be seen to have heralded a new age of animal-free meat, one which has benefited the biosphere in some ways, but also been fraught with set-backs and controversies (as described). It would seem the large CleanMeat corporations flew both too high and too low.

Notes

1. 'The CleanMeat Revolution' was the result of a five-month Residency at the Pervasive Media Studio, Bristol (UK) in 2017. An exhibition was held at Bristol's 'We The Curious', taking the form of an imagined retrospective of 21st century lab-grown meat production. The show was a museum-style display, combining artefacts from the CleanMeat industry, interpretive models, items of social history, a corrupted video lecture, and curators' notes. Together, this built up a picture of the rise and fall of lab-grown meat, providing social, cultural, and economic context for this fictional CleanMeat movement.

Author Biography

David Lisser lives and works in Newcastle upon Tyne, UK. His work concerns our future relationships with nature and technology. David has exhibited in a number of solo and group exhibitions across the UK, with residencies playing an important role in advancing his understanding of contemporary societal issues. Most recently, he has created a series of pseudo-fossils depicting possible future food production techniques ('Future Artefacts') and built a mobile seed library that is being cycled around the North-East of England setting up seed-sharing networks ('Last Ditch Attempt'). Residencies have included working with local businesses in Sunderland to investigate the ways art and business can be brought together and working alongside an industrial robotics engineer to explore the relationships between craft, skill, and automation in production line technology. David holds a degree in Fine Art from Newcastle University.

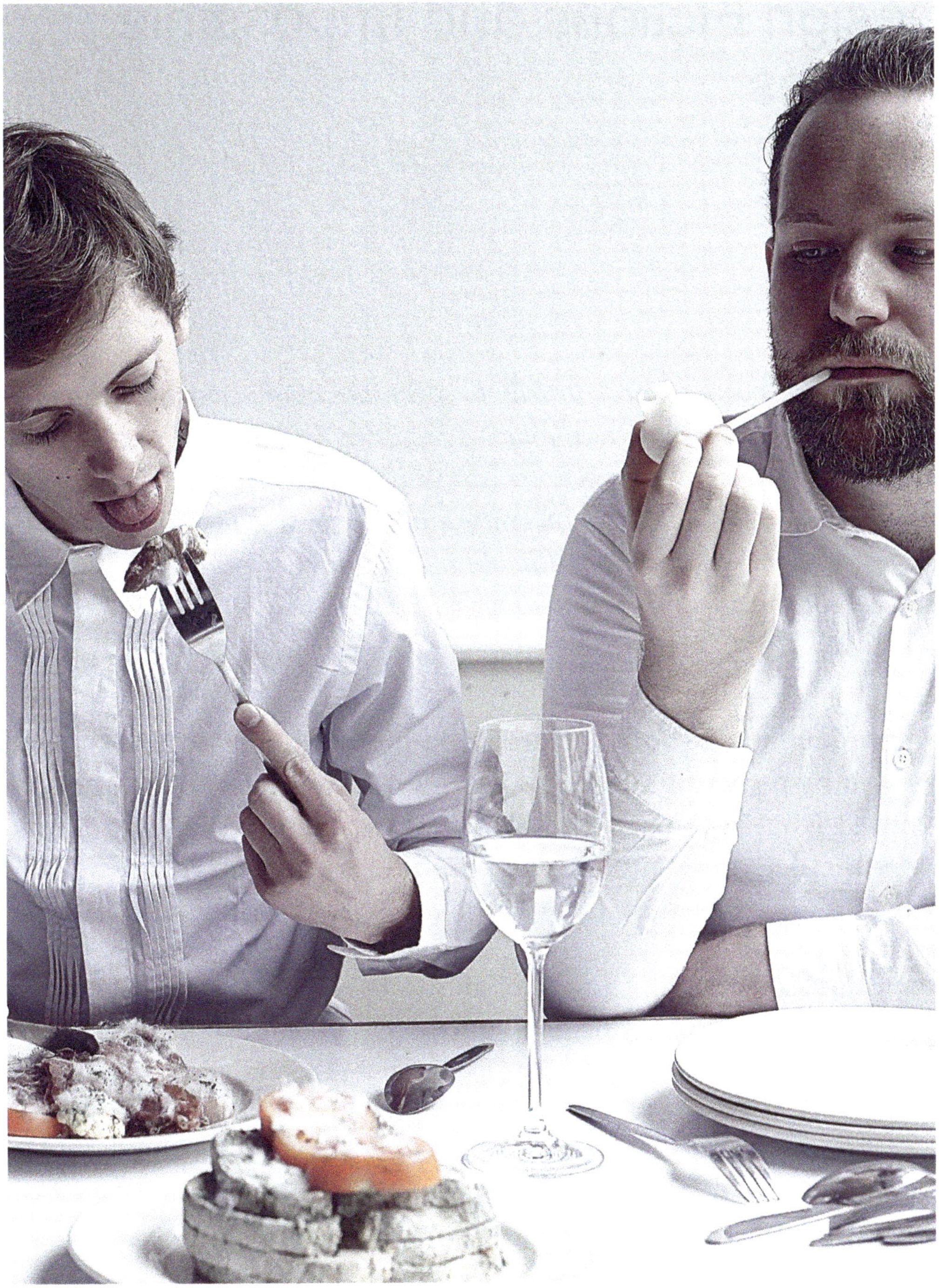

Paul Gong, Human Hyena (2014).

Design Fictions and Impossible Futures

Paul Gong

Design fictions are a powerful way to speculate about possible futures. In imagining how emerging technologies might reconfigure human, animal, and natural subjects, such fictions can be deeply provocative. In this interview, Paul Gong explores the uncomfortable prospect of tackling food waste through a form of human modification that enables a further expansion, rather than contraction, of consumer markets.

We are facing a period of increasing inequality in access to food, marked by a glaring disparity between food poverty and food excess. What is the scale of food wastage in the West, and how can speculative design help us imagine what food futures might look like?

When I was undertaking research on the Hyena Project in 2014, I read that about one-third of the food produced in the world targeted for human consumption is either lost or wasted. That is approximately 1.3 billion tons of food each year! I had thought that this statistic would be somewhat different between countries in the West and the East, but, unfortunately, it is not – our relationship to food waste seems similar. (What does differ, however, is the way we engage with food in Western and Eastern supermarkets. For example, it is unusual to have whole body parts and internal organs available in Western supermarkets, whereas it is common in the East.) An interesting subculture that has now emerged in response to this situation is Freeganism, practiced either by individuals or groups of people who go around salvaging – or in their terminology 'rescuing' – usable or edible waste from being discarded. For many, this behaviour is viewed as an effective contemporary form of foraging technique. So, with around 1.3 billion tons of food waste each year, it is clear that supermarkets, restaurants, households, etc ., are regularly filling dumpsters with 'rescuable' food items. These subcultures are interesting to me as an artist: How can these kinds of practices be developed, and what would they look like if they were pushed to extremes of scale and normalisation?

My *Human Hyena* project is an example of how I create design fictions that evoke possible and provocative futures around important topics such as food security. Here, I brought together DIYbio enthusiasts and makers to create artwork depicting future scenarios on how to tackle the increasingly serious problem we face around food wastage. What is particularly interesting about the project for me is its focus on the special ability of the hyena species to eat rotten meat without becoming sick. In trying to find out how this capacity developed, we have imagined a fictional group of humans engaging with synthetic biology technologies to create new forms of bacteria that can modify their digestive systems to be more like that of the hyena . *Human Hyenas* would be able to change themselves to adapt to the food they eat, consuming rotten food like their scavenging counterparts. Also, we were trying to explore the possibility that new food cultures might emerge around the consumption of rotten food as a way of tackling the issue of global food wastage that we are now experiencing. For the project, I have developed a series of scenario images and designed objects to present to the public. These have now been exhibited in many galleries and museums, such as the Museum aan de Stroom in Antwerp, the Museum für Kunst und Gewerbe Hamburg, and Future Gallery in Palo Alto.

Where do these possible futures sit along the timeline of our emerging food waste crisis? Are the human modifications you propose the last resort after all attempts to reduce food waste and develop a sustainable food industry have failed? Or, do you see it as a form of niche cultural innovation?

In my projects, genetic modification, or what you might term human-enhancement, is not the last resort in response to something that has failed but is more like an alternative or provocative way to get us to start thinking about how we are going to face our food future. It might not just be as simple as reducing food waste or developing new food industries; I think emerging technologies might play an important role in offering diverse and workable solutions. For example, lab-grown meat is now being researched and might very well change our food industry. Also, engaging with our food future is not a question of changing the 'natural' or 'artificial' environments in which we live, but entails changing ourselves both mentally and physically as well, so that we fit into our changing world. For me, a form of bottom-up thinking is important, meaning the use of smaller elements that we can control in detail (like individual genetic modification) to build up subsystems (like new group behaviours), and then to construct larger systems from those (such as cultural practices). I think genetic modification might be a form of niche cultural innovation in the future – one that, through the rapid emergence of new technologies, might be more easily achieved. But this will also raise serious ethical issues, with both positive and negative consequences associated with such interventions. Do we, for example, have a right to modify and change other life-forms without permission? What about animal rights? Moreover, where is the transition from modifying organisms to designing totally new life-forms? Positive outcomes might include longer life-spans and improved strength and health. On the negative side, we might face the result of being able to live for longer, with consequences for overpopulation and all that this entails.

New research now shows that, like hyenas, humans have a very low stomach pH that may reflect an earlier history of eating carrion. Whilst the hyena is, in part, a metaphor in your work, does your project in some way explore a re-convergence of natural histories – a `return to nature' that counters thousands of years of cultural and social divergence?

In 'Human Hyena', like with other projects of mine, I attempt to provoke a discussion about new relationships between humans, animals, nature, and emerging technologies. The types of discussion I try to provoke mainly focus on the evolution of life-forms in relation to the fulfilment of human needs and desires. Also, I would say that I have been trying to create through my work a nature that stands apart from, or independent of, natural histories: What can be considered natural (Nature) and what can be considered artificial (Unnature) in my work, and how they merge in 'Future Nature', is a key interest of mine. In 'The unnatural nature' (an earlier project), this presented as the difference between Nature with connotations of bio-conservation, natural selection, originality, reproduction, desire, and the unrestrained , and Unnature connoting techno-progressive, directed-evolution, mutation-intervention, change in a single generation, demand, and control. Maybe the explorations in 'Future Nature' might be understood as the dilemma between utopia and dystopia? (Although it is true that I think about the natural, I am more concerned with the relationship between Nature and Unnature. I might also describe Future Nature as a concern with 'new nature' or 'next nature' rather than the pursuit of a 'return to nature'.)

I am not sure whether this is particularly an interest common to artists today, or whether it reflects wider trends and new modes-of-thinking in society. I guess, artists today have a strong interest in the creation of novel futures and future possibilities. As is widely debated: Is evolution still a 'natural' occurrence (in the hands of long-standing, natural forces) or is it becoming 'artificial' (in the hands of man)? I think the latter might be true; I just imagine that because mankind can use technology to more precisely intervene and blur the boundary between the two states, it will. I am not saying that all artists today are 'naturally' drawn to these new forms of man-made intervention but that artists' interests in making interventions to shape the future can align with the way scientists think about the future. I think it would be great if scientists and artists thought more about the future together, sharing their knowledge to create concepts for future scenarios that are more plausible.

Your project points to new sets of relationships between modes of food production, distribution, storage, and consumption. What are some of these new relationships you envisage emerging, and to what effect?

In 'Human Hyena', I propose different fictional scenarios that connect food production and food consumption. In one, customers eat rotten food in high-class restaurants – the chef does not need to 'cook' the food but only decorate it for visual appeal. In the future, there may be many different kinds of these restaurants as we could now consume a wider palette of foods. This could also be an expression of the availability of new food resources previously unknown or underutilised. The restaurant could source its rotten food either from nature directly or from companies that collect and distribute rotten food from other sources specifically for this purpose. I imagine there might emerge a new kind of shop (maybe even simply a place or location) where we just 'acquire' food without paying for it. Moreover, the decor of dining rooms in the home or the restaurant might evolve into something quite different. Perhaps, there will be no need for kitchens with cooking facilities and refrigerators? We might just need a single space where we can store rotten food.

It is also possible that the way we consume food would change as well. For example, if we no longer have to care about food hygiene, we may have to care less about the utensils we use for eating or how we store or protect food from decay and infection. Spoiled food has different textures, tastes, and flavours to our normal fare, and this would drive changes to our preferred culinary palette and the patterns of how and when we consume food. Although we may be able to digest rotten food in the future, we will still be biologically wired to find the smell and taste of it unpalatable. There are two design elements in the project that respond to this – the Smell Transformer and the Taste Transformer, both of which use genetically modified *Synsepalum dulcificum* (miracle berry) to release enzymes that bind sensory receptors in a way that transforms all smells and tastes into sweet ones.

Your project images suggest that, in spite of a global food crisis, food culture will remain important: We see diners in your high-class restaurant retaining an elevated sense of decorum at the dinner table. How do you imagine these radically new social and cultural norms emerging?

I think that there would be strong implications for how we think about food culture. At quite a practical level, we can ask questions such as: How would we shop for, or review, good 'rotten' food? Or, what dishes might be considered romantic, bar-suitable, or family-friendly in different parts of the food service industry? We can also ask how these changes might affect our sense of cultural identity: Can rotten food be considered Kosher, Halal, or Vegetarian, for example? What about issues around 'no kill' or 'painless food' (such as eating animals who have died from natural causes or where the meat is starting to decay)? We will likely find different ways to keep food we identify with as part of our food cultures, but we may also see changes in the way we start to make, serve, or even eat traditional foods. The dishes might even combine traditional foodstuffs that we would recognise, but now in rotten form.

Here, I imagine people might maintain the way they are used to eating at first but, from time to time, challenge their own definitions around food and how it is eaten. We might even start to redefine social class in terms of food consumption (a change which has a long pedigree). If we all start to eat rotten food, and we all have food to eat, will we likely develop different relationships with food that can maintain social class distinctions. We might, for example, start to eat rotten food in fine-dining settings, with certain foods becoming a new symbol for a high culture associated with particular forms of decoration, preparation, and hygiene standards. Might, for example, the most rotten food – the food that is hardest to come by and digest – become the most valued and sought after as a class-distinguishing feature? Perhaps, the longer the food decays, the greater its flavour and appeal will become!

Returning to the proposed intervention itself: The relationship between the pH and microbial diversity of our stomachs, and how a balance can be achieved between healthy and pathogenic elements in the gut microbiome, is complex. Striking a change in this balance in response to new food pressures will be an unpredictable and potentially dangerous process. How do you envisage this act of DIYbio unfolding?

In my own work, I am an artist assuming the role of a DIY-biologist, so what I describe is more of a speculative process that makes use of fictional scenarios. I think that transhumanists, DIYbio enthusiasts, as well as makers could certainly be a part, if not the centre, of such a revolution at the frontiers of human modification. I think that I have shown this to a certain degree in 'Human Hyena' as this has proven a subject with appeal to all these communities, as well as evolutionary biologists, gastrointestinal researchers, and geneticists. So, in spite of the strong citizen science aspect to this work, there is a need for scientists and other professionals or experts to be involved to ensure these DIY approaches work effectively. For example, it would be important to include synthetic biologists and microbiologists as project consultants, so a DIY team could gain access to appropriate methods and training and to ensure that our work runs correctly in regard to health and safety concerns. Other forms of disciplinary expertise that have, are, or will be important to the DIY community include psychologists – who would be needed to analyse the mental states of those undergoing modification – and evolutionary biologists – who could collaborate together with psychologists to discuss which developmental routes are more mutually beneficial to our physical and mental condition.

Yes, we would still need to follow the logic of Science, and this would involve lots of research. But the DIY community also needs more than just disciplinary professionals: It needs people who can also work at new levels of interdisciplinarity in order to truly create new knowledge and understanding through collaboration. Disciplinary experts could collaborate together to tackle different layers of issues raised in the creation of blueprints for what I might call the 'Human Hyena' revolution. It is the fact that we appear to make this 'possible future' plausible, but also fantastical, that might make it all one day – perhaps – even possible. We are seeding the ideas, and, together, we might make it a reality. Several research institutes have even shown interest in the 'Human Hyena' project, and, in our discussions with them, they mentioned to us that the project offers a way to re-think the many inherited relationships between humans and food. For example, the 'Institute For The Future' created an event in 2015 as part of their Ten-Year Forecast called 'Café Hyène: A Speculative Dining Experience in 2025' in which a chef was brought in to create suitable menus for audiences to encounter this possible future.

Would it be fair to say that there is a strong case to be made for our `becoming media' in your work, i.e., the human body becoming a target for designerly interventions that convey a new message around self-determination and adaptation?

The concept behind the 'Human Hyena' project could be expressed just like that. The main purpose of this project is to offer an audience the opportunity to imagine the possibilities of unknown futures and alternative worlds that might be out there. This is done through confronting them with technologies that they will know from the news as playing a part in the new revolution on evolution. People see the use of biotechnologies in genetic modification, and so see that we are clearly changing the world around us and making it different. In 'Human Hyena', the body – and, therefore, the body of the work's audience – is the medium for those future possibilities. The possibilities for making these changes are now here, but they are certainly not all for the better. Above all, we need to get people to think deeper about the utopian and dystopian elements of these scenarios. By imagining or even witnessing the behaviours of the 'Human Hyena' (through the presentation of one such future in which artists portray themselves as being part of the work through ingestion), the audience considers the possibility of doing the same when confronted with the same scenario.

I think of this project as expressing a timeline in relation to the speculative scenarios for possible future applications involving advances in biotechnology. This timeline expresses the present, the past, and a prediction of the future. The first two facets help us reflect on our present situation today and how we got here, readily acknowledging current technological advances. The third lets us imagine the different possibilities that lie beyond our current capabilities. I see the audience as being critical in the expression of this timeline, with the work aiding them to think outside of the limits imposed by our current reality in order to reach an understanding of what might be a looming food crisis. Through this, our minds are opened up with new ideas that embrace different possibilities for a future that might avoid it.

Author Biography

Paul Gong is a speculative designer, artist, and curator working in Taiwan. He uses scientific research as an inspiration to create design fictions that evoke possible and provocative future scenarios. He holds a BA degree in Industrial Design from the Chang Gung University in Taipei, and an MA degree in Design Interactions from the Royal College of Art in London. His work has been exhibited at MAS (Museum aan de Stroom) in Antwerp, Museum für Kunst und Gewerbe Hamburg, the Taiwan Design Museum and Yiri Arts in Taipei, USC 5D Institute in Los Angeles, and Future Gallery (in Palo Alto, London, and Guangzhou). As well as being an independent designer and artist of Ouroboros – Organic Organisms of O (Artist Collective), he is also a part-time tutor of the Department of Industrial Design at Chang Gung University and a part-time lecturer of the Department of New media Art at Taipei National University of the Arts. He was awarded the Next Art Tainan Award in 2018. More on his work can be found at https://www.paulgong.co.uk/.

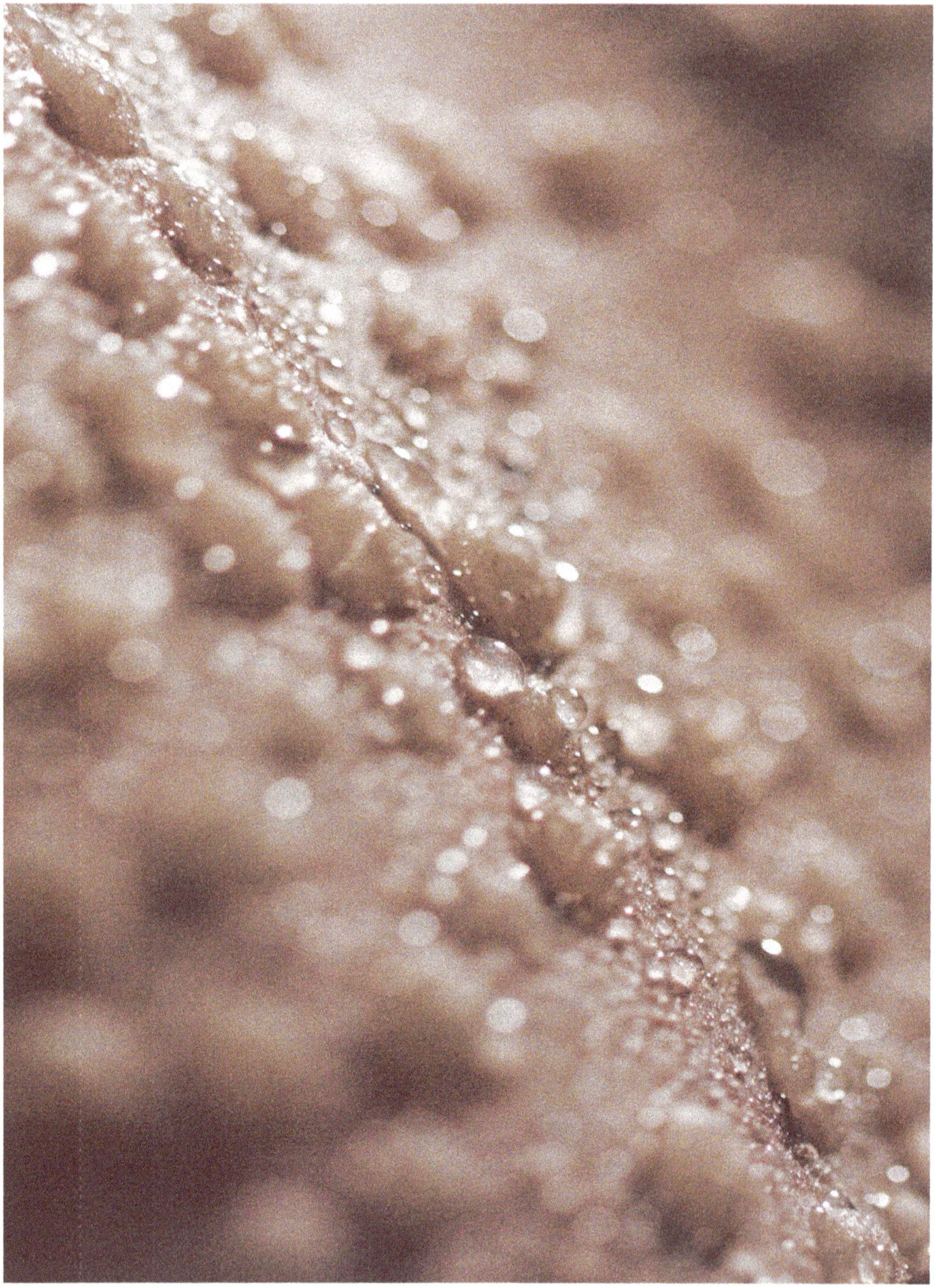

Nestor Pestana, from After Information series, The Exudaters (2015).

Malleable Bodies: Life Beyond Utilitarianism

Nestor Pestana

We have modified ecosystems around us to control the means of food production over millennia. In how far we understand this in terms of relationships between ourselves and other forms of life is a question each generation asks anew. In this interview, Nestor Pestana asks whether our conception of ecosystems as a source of nourishment that exists external to our own bodies is now an idea ripe for challenge.

When we think about food production, we inevitably make reference to forms of interdependence between different living, non-living, and technological systems. Where have you imagined in your work a different site for those interactions to lie?

The human body is highly malleable, and new and emerging technologies are now allowing us to make more significant and profound interventions than ever before. My project – 'The Exudaters' – is a conceptual piece exploring how we might modify our biological systems to attain our most complex desires. I was particularly interested in using design approaches to take advantage of the symbiotic relationships that we have with microorganisms, such as bacteria, and looking at the human body as a source of production – a little bit like a farm (our bacteria the crops and animals, our own flesh the land). During my research, I learned that there were many libertarian communities in the 19th century Essex, UK. Whilst some were self-sufficient, others promised salvation to juvenile delinquents through labour or a way of life that ensured their inhabitants' entrance through the gates of heaven. They were social and utopian experiments: escapists, fourierists, and owenists. They also did not last very long.

The narrative of 'The Exudaters' is modelled on these Utopian communities of Essex, depicting a biohacker community living in isolation from an industrial and materialistic world (and perhaps even sharing a similar fate). The project's focus was not so much on whether the Exudaters succeeded, but rather on how biotechnology might allow biohackers to pursue similar libertarian goals – albeit to more extreme ends. Perceiving their bodies as a complex combination of living agents that can be enhanced or maximis ed to produce all they need for survival, they essentially push the boundaries of the probiotics industry through biotechnological interventions. Not only do they introduce new bacteria into their bodies (such as *Synechococcus elongatus*) and genetically enhance them to produce nutrients from sweat, but, most importantly, they design and bioengineer the ecosystem for the bacteria to operate in. The result is sweat glands in the skin that have been modified to serve three functions: The first is to contain the sweat produced by the body; the second is to host the bacteria which feed on sweat and excrete nutrients; the third is to absorb the nutrients produced by the bacteria into the bloodstream. These bioengineered sweat glands become, effectively, tiny digestive systems. More than trying to imagine a self-sufficient human being, however, I was interested in how technologies can push us beyond utilitarianism, i.e., how they might be used to fulfil our own ideals. The threat of pain or disease come to mind when we think about the transformation of living human matter in the way discussed. 'The Exudaters' depicts a world where we have surpassed these fears and are now able to understand matter beyond pain, and aesthetics beyond disease.

You address the condition of post-humanism through your work – a speculative endeavour to imagine future human capacities expanded through new technologies. What does post-humanism mean to you, and what themes within this condition strike you as most interesting?

Post-humanism embraces the idea of a human being defined not only by a biological body but also the technologies that are produced at a given place and time. For this reason, and through the lens of post-humanists, we are in a constant state of definition that depends on our contemporary technologies. For example, in the current informational era (one governed by informational technologies), data is what primarily defines us; we have the urge to understand ourselves through it. Take the human genome project – a pure translation of our materiality into a series of codes that ultimately hold the promise that, one day, we will take evolution fully into our own hands. But this post-human definition cannot be detached from the socio-cultural, political, and ecological contexts of the time and place in which it exists. They inform how we see ourselves and the world around us, and how our obsessions, frustrations, desires, and aspirations manifest in relation to 'being human'. These are interesting subjects to explore for any artist or designer.

In my work, I have been exploring speculative scenarios where we modify our bodies through emerging technologies – a very post-humanist subject. These modifications, however, are problematic in our real world: Our bodies adapt on an evolutionary timescale, and as we introduce more 'novelties' into them, we are forcing rapid change without time for our bodies to adjust. (We also introduce body and gene modifications that become sites for commercial interest and activity.) If we are to tackle this post-humanist world responsibly, I think we have first to seriously address the social, political, and ecological inequalities that characterise life today. I do not claim to have answers for how we achieve this, but evolutionary theorist Bret Weinstein presents an interesting view: He states that a tremendous amount of what we are is stored not in our genomes but in a cultural layer that is passed on outside of genes. This layer is vastly more flexible and easier to grasp than our genomes, with the possibility of providing mechanisms to help us increase the capacity of our minds to address such concerns and clearly identify ways to solve them, so enabling us to move forward as a species. That being said, I think that whatever methods we use to create the right kind of environment for an exploration of our bodies through technology, working with sensitivity to the dangers of utopian idealism will be critical.

`The Exudaters' is a work that seems to point away from the realm of fiction. Not only is it a piece that gets figuratively under your skin, but your consultation with a scientific team suggests a kind of `actionable future'. How have people responded to, or been drawn into, your work?

The aesthetic language of my work plays a fundamental role in drawing people's attention to my pieces, inviting viewers to dive deeper into their conceptual and scientific layers. I see an eclectic range of responses according to the project in question, but there is a common theme of 'shock' (especially in my 'After Information' series), which I think is caused by people's confrontation with another human being that has a modified and unusual appearance. Perhaps, people feel the work in their own bodies too. In 'The Exudaters', for example, the human body becomes exposed in a new way as a malleable material – one transformed through a visual language borrowed from the realm of human disease, namely blisters. I am conscious that this might be a little disturbing for some, but I think that developments in biotechnology are going to inform our aesthetic models of the body in a much deeper way than they do today.

The film produced for 'The Exudaters' is quite abstract, partly because I wanted to capture a scenario that is highly speculative in nature – in line with the utopian, self-sufficient values held by the community in the film. I wanted people to perceive the piece as something that is not part of this world, and perhaps never will be (which, in a way, is what utopias are really all about). But this abstraction, along with its strong visceral aesthetics, creates a sort of tension that directs people either to look for a more scientific explanation of the piece or to remain in a state of aesthetic exploration: On the one hand, there's an audience of people with a scientific background that tend to engage with the research behind the project (or at least they seem to fully understand this aspect and, perhaps, are looking to see how other people, such as artists and designers, are envisioning a translation of this scientific knowledge through their work); on the other hand, there is an audience with a greater distance from Science that seems to be curious about the opportunities and threats arising from scientific advances that will shape our future lives – I suspect that they are also the ones in shock.

In `Only Information' (Post-heaven), you explore the possibility of human states that are, in contrast to `The Exudaters', digital and hyper-connected. Where do these two projects converge or diverge as part of your ongoing exploration of the post-humanist condition?

Both projects are about the transcendence of matter, imagining scenarios in which we fully control it rather than being subjugated by it. Here, we take hold of our own evolutionary paths, manipulating our bodies beyond their current human form and limitations by means of science and technology. But the projects also diverge in many ways, principally in the type of technology that is being explored (informational versus biotechnological) and the way matter is conceptually and philosophically perceived through the lens of those technologies.'Post-heaven' explores the desire, enabled through informational technologies, to get rid of the body and so become pure information; this is the state that primarily defines who we are as seen through the lens of this technology. Materiality, and hence the human body, is secondary. If we were to become pure information, we would exist in a universe of material abstraction and mathematical formulae, interconnected in a dimension of nothingness. This is something we are not able to experience or understand because the human condition is limited to materiality – each of us connected by a culture but separated by matter. Perhaps, this is what death is all about? Stephen Wolfram, and most recently Elon Musk, has suggested that we might live in a computational universe, and matter is just a simulation that allows us to experience reality. In this sense, informational technologies are not promising us the power to translate ourselves into different material formats – this is something that is already happening. In contrast, 'The Exudaters' explores the direct manipulation of our material reality: It imagines a more complex, rapid iteration of the human form as enabled through advances in biotechnology. In this way, it addresses how human beings are made of complex interactions of different types of living organisms, such as cells and bacteria, all playing a crucial role in constructing our experiences in the world. It also focuses on the importance that the body (and all its constituent organisms) has in the construction of our identities, and the role technologies play in providing us with tools to further express ideological discourses through matter. The project is ultimately a celebration of reality in its different forms.

By pushing the boundaries of the human body through these two technological approaches, I was trying to understand the strengths and weaknesses of each position, a route to answering the question of why we are so enthralled by technology. The answer I think is a very simple one – a desire for growth. We seem to seek in technology the solutions to global economic growth and sustainability, as well as the solutions to our own most personal growth and development. But technologies always have unpredictable consequences, and they often are not the right solution to a given problem: It is these issues that I am committed to exploring through my practice.

Both projects envisage new types of relationship between our biological selves and emerging technological, social, and market economic forces. What are some of these interactions you have been considering, and are these different forces now becoming more inextricably linked?

New technological developments exert both positive and negative impact on our social and economic landscape: Positive in that they can be a motor driving society forward, for example, with regard to scalable medical advances, but also negative in how they generate inequalities, ultimately around who has access or control over new technologies. These are complex issues with overlapping and blurry boundaries. I find the social dynamics raised by the biohacker communities particularly interesting in that they have taken technological developments into their own hands, learned how to manipulate them, and even successfully incorporated them into their own bodies. In other words, they do not need experts to make these procedures because they have gathered a level of expertise themselves.

This triggers a series of ethical and legal concerns, but it also raises fundamental questions about body ownership. Germany, for example, has banned such biohacker practices. To what extent do others have the right to dictate what we can do to our own bodies? Although I am of the opinion that we each should have the ultimate say over our bodies, we should not ignore that such experiments might lead to the kind of injuries requiring medical assistance ; if publicly funded health care services are brought into play, then taxpayers will be indirectly contributing to such experiments, even without their consent. We will need new types of regulation over emerging technologies and technological practices to prevent them from being abused. A stronger engagement around the ethics of technology is also going to be needed. What if these new approaches are used to create bio-weapons? As John Gray puts it in his book 'Straw Dogs' , 'New technologies of mass destruction are cheap; the knowledge they embody is free'.

Do the two post-human conditions you describe exist as alternative states, or might they sit side-by-side in diametric opposition within the same lived reality? Put another way, when does living as a disembodied brain or a body-dependent Exudater become a matter of choice (or last resort)?

These two post-human conditions (concerning the biological and informational) are to some extent antagonistic: One is about wetware and biological matter, the other about hardware/software and the absence of biological matter. Taking the post-humanism principle that our technological landscapes play a role in defining us, one could imagine the existence of two types of humans in the future: One smelly and biologically enhanced, the other odourless and living in a series of microchips. Projecting this idea further – imagining a course in which transhumanists were to achieve their end goal of immortality – we would soon witness a highly unbalanced distribution of technological control. Perhaps, in this extreme dystopian scenario, having a physical body might even become a luxury, the joys of experiencing reality (including dying) reserved only for a technological elite, whilst the rest of the post-human population would live in an immaterial, labour-led, and death-free world designed to sustain the material world.

Although the main goal of many transhumanists is to become immortal, I imagine that if they were to achieve such a state, they would soon realise the value of dying, if only to put an end to one life phase in order to start afresh with another. (We could, of course, imagine in such a scenario that artificial systems able to mimic death might be developed, again with access restricted to some and not others.) These speculations might seem a little far-fetched, but they reflect common concerns for our current social and technological landscape (that is, the role of information in sustaining the material world and generating inequalities through the way technologies are designed and controlled). Sadly, I do not think we currently have the right cultural and social frameworks in place to develop such transhumanist ideas responsibly without moving towards dystopian scenarios such as those described. Perhaps, we need first to develop our capacity to address the many social and cultural injustices that we face today.

Science fiction has been a key stimulus for your work, offering original thought experiments around the science of post-humanism. In your eyes, what role does science fiction play in how to envisage the complex challenges and opportunities that arise from the use of new technologies?

The things that we produce through emerging technologies are often confined to labs and other controlled environments, isolated from the rest of the world by safety and containment requirements. As these new artefacts and objects get developed, we will start to see the move from these controlled environments to more complex, diverse ecosystems. This is when things become interesting and potentially messy. We have already witnessed how disruptive the introduction of a new element (biological or technological) can be to a foreign ecosystem. For example, when we first brought the car into our lives ('everyday life' as a complex but tightly balanced ecosystem), we also introduced car crashes and pollutants released through combustion. Such ecosystem interventions often have unforeseen consequences, especially when the 'thing' introduced has been developed in isolation from the rest of the world. Furthermore, it is just impossible to determine or predict how all components of an ecosystem will react to a new element added into it.

So, we can only speculate, and this is when science fiction can actually play an important role. It can be used as a tool to explore such interventions by, for example, examining the potential impact of a technological product in a given ecosystem, and so help us imagine what sort of new dynamics and consequences might arise. As nothing operates in isolation in the real world, this can provide us with a more holistic understanding of new technological interventions being developed in controlled environments but destined for 'release' into a wider ecosystem.

As new interfaces between art and design, biotechnology, and economics emerge, we can surely expect the parallel development of new areas of crossover expertise. What is the role of the artist today in not only exploring speculative scenarios but also collaboratively testing the boundaries between science fact and science fiction?

The process of generating speculative scenarios often requires working in a multidisciplinary team, one involving like-minded individuals able to raise questions from within their field of expertise, and ready to say 'how things might look' in the world that is being imagined. I think it is the richness of these collaborations that lays the value and relevance of a sci-fi project. As important questions are explored together, creative outputs emerge that can then shape how ideas develop out in the real world. Collaborations are becoming increasingly important in defining the role of the artist and designer today, especially those interested in geopolitics, philosophy, science, and technology. These subjects are too complex to be dealt with alone. The role of the artist is to find a way to effectively translate these dialogues into something tangible and meaningful to themselves and the public they are trying to reach, promote creative and critical thinking, and both share and exchange knowledge along the way.

Although the formulation of a speculative project might be, to some extent, similar to those addressing real-world utilitarian concerns, the pressures and anxieties are, of course, very different: The first is wholly conceptual, the second practical, and this is how they should rightly be distinguished in any approach. Ultimately, both are trying to respond to a kind of reality (one that is imagined versus one that actually exists in the world), which is why they tend to inform each other so strongly. In my experience, scientists and other experts have shown a healthy interest in fostering such crossovers that involve creatives and fictional practitioners in their investigations; it is, for them, an opportunity to think differently about what they do, which might, on the one hand, enrich their research, and, on the other, help them translate their work into formats suitable for public engagement. The reappraisal and reimagining involved in speculative projects does require a high level of research and partnership capability, especially if the project pushes the boundaries of plausibility and predictability.

Of course, collaborations do not always run smoothly, and there is no formula for how they should be conducted (or who should be involved). Each project is a different journey with its own needs and specifications, so collaborations need to be tailored accordingly. I normally follow a loose plan to start with, one based strongly on both research and intuition: First, I test the project with people I think could bring relevant insight to the project (often before inviting them formally to take part); then things start to happen more naturally when we are all on the same page, working with the same dedication and energy. This might all sound very generic, but I really do not believe that a successful collaboration can flourish in an environment devoid of these characteristics.

Author Biography

Nestor Pestana is a speculative designer and multimedia artist based in London. A core focus of his work is the post-human condition, one in which the human body is understood as malleable material for modification through innovative emerging technologies. His work is collaborative and multidisciplinary, finding expression in a variety of media that includes props, films, animations, interactive installations, workshops, and performances. He holds a bachelor's degree (Hons.) in Design from the University of Aveiro and a master's degree in Design Interactions from the Royal College of Art in London. Exhibition highlights include 'Paths to Utopia: A Night School on Anarres' at Somerset House in London, 'Bio-Art Seoul 2015: Abundance of Life', and the Swiss Pavilion's 'School of Tomorrow' at the Venice Architecture Biennale. His work is in the Wellcome Trust Collection. More on Nestor's work can be found at http://nestorpestana.co.uk/

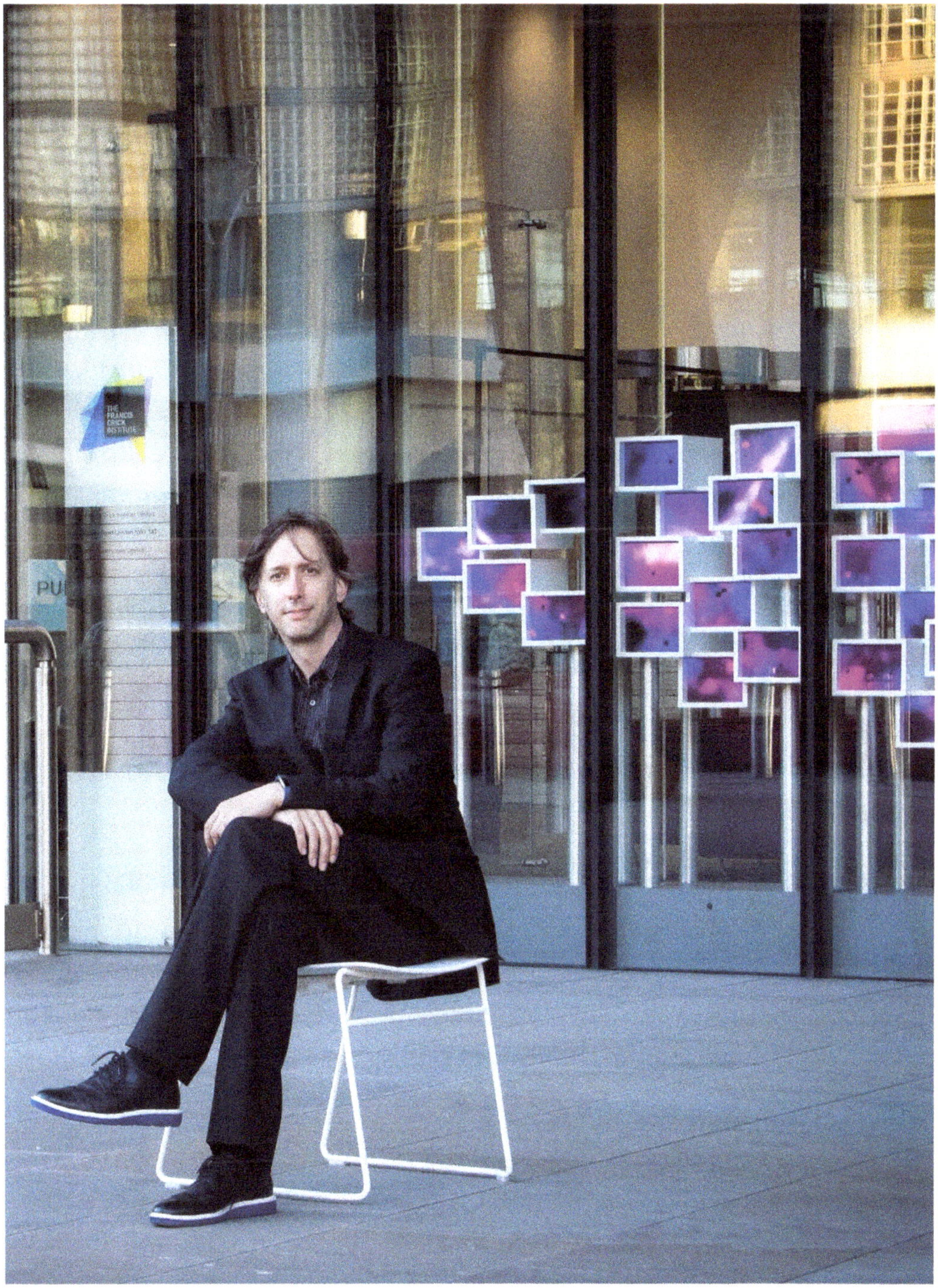

Alex May, Flow State (2018).

From Petri Dish to Big Data

Alex May

As we begin to understand the human condition in terms of a `multi organism', we need to ask more about the interactions that sustain life between the multiple living bodies involved. These are relationships that generate data, sustain information exchange, and build a shared heritage. In this interview, Alex May explores how his work as a creative technologist can open up a very human engagement with the human condition.

Computer interaction techniques can be a powerful way to refract living processes into different informational streams, so bringing them `to life' in a different way. With works such as `Sequence' and `The Human Super Organism', how are these techniques changing the nature of our interactions with bacterial life?

In both 'Sequence' and 'The Human Super Organism', the interactive element encourages visitors to explore, in an engaging and educational way, aspects of the complex relationship we as humans have with our bacterial ecosystem; in this way, they bring new knowledge from cutting edge research and bioinformatic techniques out of the lab and into the gallery.

'Sequence' is a work that offers a VR-based experience to visitors, leading them through the physical processes and healthcare implications of whole genome sequencing. Its starting point was work by the artist Anna Dumitriu on *Staphylococcus aureus* bacteria that she has been culturing from her own body since 2010. From 2014 to 2015, Anna worked in collaboration with the Royal Sussex County Hospital (Brighton, UK) and the 'Modernising

Medical Microbiology' project (led by the University of Oxford, UK) to undertake DNA sample extraction and preparation, load and operate the whole genome sequencing machine, and process the raw data generated using bioinformatics software to arrive at a DNA sequence for the bacterial samples. To create the piece 'Sequence', we took the raw and processed data from the sequencing process and developed a VR environment around it using my bespoke software, Fugio. The VR setting allows the visitor to fly through the extracted and reconstructed data, letting them come face- to-face with the 'big data' of a single ring-shaped bacterial genome (2.4 million DNA base pairs). The project was supported by Arts Council England, the Royal College of Pathologists, and Oxford University's Knowledge Exchange seed fund, and it was premiered at the Victoria & Albert Museum as part of the London Digital Design Weekend in 2015.

The 'Human Super Organism' is an interactive digital installation that reveals the abundance and diversity of our commensal bacterial ecosystem . Similar to 'Sequence', which relied on cultivating bacteria from our own bodies, Anna Dumitriu and I cultured our own skin flora onto homemade agar plates and filmed them in a custom camera enclosure. The method we developed to do this involved making high-resolution time-lapse videos, which were then cut up to capture individual bacterial cultures growing within sections of the agar plate; these were then used as the source imagery of the work. To interact with the work, visitors place their hands on a large projection screen – acting as a virtual petri dish – for a few seconds . On the screen, the silhouette of the visitors' hands appear filled with bacteria, these made from the cut-up video sections described, composited in real-time using Fugio, and then projection mapped onto the screen; once activated, the bacteria then go through a life cycle of growing and dying off. Commissioned by Eden Project with support from the Wellcome Trust, this work was based on previous projects commissioned by CineKid Festival (NL) and the Wellcome Collection.

Big data offers us the promise of previously unattainable levels of detail relating to life processes. Yet, in its abstraction and sheer quantity, it lacks the very *singular coherence* we attribute to life. Are you using arts practices around data visualisation to bring a sense of unity back into big data?

The scientific process of visualisation consists of preparing an optimised selection of information that is reliably reproducible across a particular type of source data sets. This is done in order to prepare the checked data for further study and classification by the human observer . Conversely, to work with even a single set of big data in its complete, raw form is an experiential proposition. One is faced with a scale of information that has no meaningful start or end point. The ring-shaped genome of the bacteria in 'Sequence' is a good example of this; the experience of approaching a Big Data set feels like standing under the stars of the Milky Way on a dark night. On a comparative physical scale, and in terms of temporal existence, we can truly sense the magnitude of that information. In looking beyond the organism into the genome, we must engage with abstractions and data that are hard to interpret or make sense of. There are further levels beyond that of chemistry, physics, and quantum mechanics that we cannot feel or smell either ; so the challenge is making some kind of meaningful link to what we understand in the everyday. The visualisations that we present in 'Sequence' and 'The Human Super Organism' were created as much with the intention of bringing the visitor face-to-face with an experience of the magnitude of such data as with confronting them with the meaning that might possibly be derived from that data.

Part of that 'new meaning' is a reappraisal of what it means to see ourselves as individual human beings versus a part of a wider system of organisms and relationships. For example, bacteria, historically speaking, have been understood as separate from us, as something either 'good' or 'bad'. However, through the public dissemination of science, we are all now learning just how deeply integrated our physical and psychological existence is with these minute life-forms; this is quite a switch from our dominant human-centric world-view. Artworks like 'The Human Super Organism' aim to introduce people to aspects of these discoveries. The work encourages people to learn to accept the fact that we are literally covered, inside and out, with commensal bacteria. In this way, our artworks are more than just an engagement with the products of scientific enquiry and big data; they open up a unique enquiry for each individual that interacts with the works as they explore questions about what it means to be human in a world of 'super organisms' that have shared heritage, engage in symbiotic relationships, and so on.

Artworks based on living materials can engage us directly with life's generative and unpredictable nature. Can digital techniques (such as projection mapping) bring the simulation and re-presentation of living processes into the same kind of close proximity that a `living encounter' can offer?

The line between what is digital and what is not continues to evolve, with advances in the fields of visual and audio technologies (over other senses like touch, smell, and taste) best known and more publicly available. This interests me in relation to the presentation of 'living systems' in that whilst we can present some kind of simulation (such as in 'Super Organism', where visitors press themselves against a projection screen that looks like a giant petri dish, and they see colonies of bacteria grow in the shape of their body), it is the physical interaction and involvement that makes the experience work. There is a visceral and experiential moment where you are forced to be in your body, feeling it pressed against a physical object before you stand back to visually evaluate the results. This extension into just one additional sense (touch) brings an important extra interactive dimension to the experience of the work and the living matter represented in it. It is the innate ability of digital technology to i) respond to such inputs in a non-trivial way; ii) respond in real-time to the nuance of the physicality of the participants; and iii) present a narrative and guide the participant through it without being didactic that holds the promise of a wholly immersive experience – one that does n ot need to address all senses unless it is enriching to do so.

In my experience of researching artworks based on lab works, there is a rich palette of aromas that vary from room to room based on what processes and life-forms are being worked with. They are not always pleasant, but they are part of living systems and so capture another way for us to understand or interact with them. Scientists are rarely aware of these aromas (having become desensitised to them through repeated exposure over many years), but for a first-time visitor, they can knock you back. While I am not suggesting that artworks that lack an olfactory or somatic component are less able to convey a meaningful experience or message, these other sensory experiences remind me of how living things inhabit a richer spectrum than that generally encountered through interactive artworks dominated by a visual sensory component. To explore this field further, we are currently working on a new interactive robot with a 'nose' – one that can smell specific compounds in the environment and physically react to them.

There seems to be an interesting parallel between the endless, shifting grounds of scientific knowledge and the fleeting nature of digital practices. Is there a need to preserve the digital works and immersive environments of our age if we are to understand in the future how we got there?

It is an exciting time to be working with creative technologies because they give me the tools I need to integrate with a wide range of developments in a countless number of fields. It can also provide a 'common tongue' when talking to scientists and bioinformaticians, where applied knowledge of certain algorithms and techniques is relevant for many areas of scientific research. (For example, the F ast Fourier transform, or FFT, that translates signals between the time and frequency domains, is something I use a lot for real-time musical analysis, but it is also something commonly used in the sciences). On more than one occasion, I have been able to hold much more in-depth discussions with scientists after telling them about the technologies I use every day as part of my art practice.

There are some interesting differences, perhaps, in how new tools are superseded in different fields. As better tools are developed, and valid information comes to light through their use, we can, generally speaking, safely leave old tools behind. Scientific researchers would hardly ever choose to use antiquated technology, particularly if there was a better solution available to them at that time. There is, however, an experiential quality in the digital realm that has proven itself desirable to preserve. For example, the 'MAME' project has developed a computer program that emulates old arcade machines, making thousands of old arcade games playable once more (many of which are still fun to play and elicit a joyful, deep reminiscence). I am very much of the opinion that those who create digital artworks should pay some thought to the proposition of preservation so that, in the future, people can fully experience an artist's original vision of their work, rather than just reading documentation about it. The preservation of digital artworks is something that I have spent the past fifteen years thinking about and working on. It is a complex area, and, above all, it requires a good grasp on which technologies provide the possibilities for preservation and which do not.

Developing this further, what are some of the key current technological and cultural shifts in the use of digital practices that enable preservation of digital works or introduce difficulties into preservation activities?

We are seeing a shift of technological control back towards centralised servers and services; originally an issue of physical necessity (computers of any power were large and expensive), this has now become a necessity by dint of how many people use computers daily around the world. Companies like Apple and Microsoft, who spent years offering more desktop features and power, recognise the vast majority of their consumer/office market want to do relatively few things (web browsing, email, photos, etc .) and so do not need powerful computers. By stripping out lesser-used features and shifting others online, these companies have fewer user-support issues or software bugs (potentially) to worry about; and if they break some digital artworks and frustrate a few artists here and there, who cares? While a large number of computer users were once technically savvy enthusiasts and early adopting creatives, they now represent a small part of the market; powerful computing devices have become ubiquitous facilitators in all of our working and social lives. But at the same time, we have seen the growth of platforms such as Arduino and Raspberry Pi. While not entirely open source, these self-contained computers are cheap to make, cheap to replace, and powerful enough to run open-source operating systems. Although they are not always suitable for projects that require vast amounts of raw computing and graphics power (where clustering or other innovative solutions will be needed), these qualities – durability, mass production, and open design – do make them a good choice for developing and preserving works.

Seen from another perspective, however, creating new work on closed platforms affords certain advantages, such as being able to use tools not yet available in open-source form; this may be vital when artistically responding to contemporary developments and conversations around digital concerns (although one must accept that such works are built on shifting ground with no guarantee of longevity or support). Along these lines, I consider emulation (and the ability to be emulated) a key property of technology primed for preservation activities. Operating systems such as Linux (and its many variants) offer the most promise due to their open-source policies. Windows has traditionally been a good next choice as it is relatively simple to run under VirtualBox, Bootcamp, and other emulation layers, although the direction in which Microsoft is taking Windows (more towards a managed system) may require a re-evaluation of whether it remains fit for this purpose. Apple's macOS (a much loved operating system) is much harder to emulate (although possible using Hackintosh systems) and actively fights any attempt to run it on unsupported hardware (i.e., those not made by Apple). It is telling, and inconvenient, that the most closed system is a product of the first company to be worth over one trillion dollars.

Much of your work is made possible through funding and support from UK university researchers. You are currently working on a commission for the Francis Crick Institute in London. How do these connections reflect the changing landscape of our engagement with scientific knowledge and those who can shape in it?

Working with the Francis Crick Institute has been a fascinating exploration into the boundaries and crossovers between different levels of public/private space. There is a proactive desire to bring the public into the building to meet scientists, learn about the work they do, and find a platform for discussing their concerns with the kinds of issues scientists at the Crick are working on. The scientists I had the pleasure of working with on the project have been very open and relaxed about being involved with the production of an artwork. They recognise that the piece is not an exercise in science education, but rather an opportunity to reveal research processes and visualise information that the public would not normally be able to witness, i.e., creating an aesthetic exploration of scientific work where anyone from the public can ask questions about what they are seeing and why what they are seeing behaves like it does – the kind of questions that are posed by scientists every day.

From my experiences of developing that work for the Crick Institute, I feel that there is a real opportunity to build on the potential for connecting across disciplines and creating new flows of information between fields. Basically, I think that we need to focus more on the commonality of our humanity and our innate shared curiosity . It would not open all doors for new partnerships, but the better able projects are to cross disciplinary boundaries, the more we will benefit from the rich insights and experiences they can offer. This is true both in terms of knowledge created and in how those exposed to (or involved in) such work will think about their practice or methods in the longer term. In this way, we will see a more seamless use of creativity, and a wider range of opportunities for working together, opening up.

There is, as such, a growing recognition that artists can bring unique and unexpected insights into this 'common curiosity' that drives humanity to strive, to explore, and to learn . How these partnerships can be best supported is an evolving question. My personal preference is for artists to work with institutions on long-term art projects, not as part of a scheme that tries to instrumentalise them for the purpose of generating new innovations. Of course, innovation may happen as a by-product of the work the artists are undertaking (or the environment in which he or she is operating), but the principle purpose of such collaborations should be for the artist to create the best possible art. However, for that to happen, all doors must be open for an artist's curiosity, and there must be ample time for new work to be conducted and sufficient support given to them.

Author Biography

Alex May is a British digital artist working with a range of media, including code, video mapping, performance, interactive installations, VR, photogrammetry, algorithmic photography, and robotics. His work concerns the human condition in a hyper-connected, software mediated, politically and environmentally unstable world. Alex has exhibited internationally, including at the Tate Modern and the Wellcome Collection (UK), Ars Electronica (AT), LABoral (ES), the Museum of Contemporary Art in Caracas, the Science Gallery in Dublin, and at university institutes in the US and Canada. He has artwork in the permanent collections of the Francis Crick Institute and Eden Project (UK). Alex is currently a Visiting Research Fellow and Artist-in-Residence with the Computer Science Department of the University of Hertfordshire, UK, and a sessional lecturer on the Digital Media Arts MA at the University of Brighton, UK. More on Alex's work can be found at https://www.alexmayarts.co.uk/

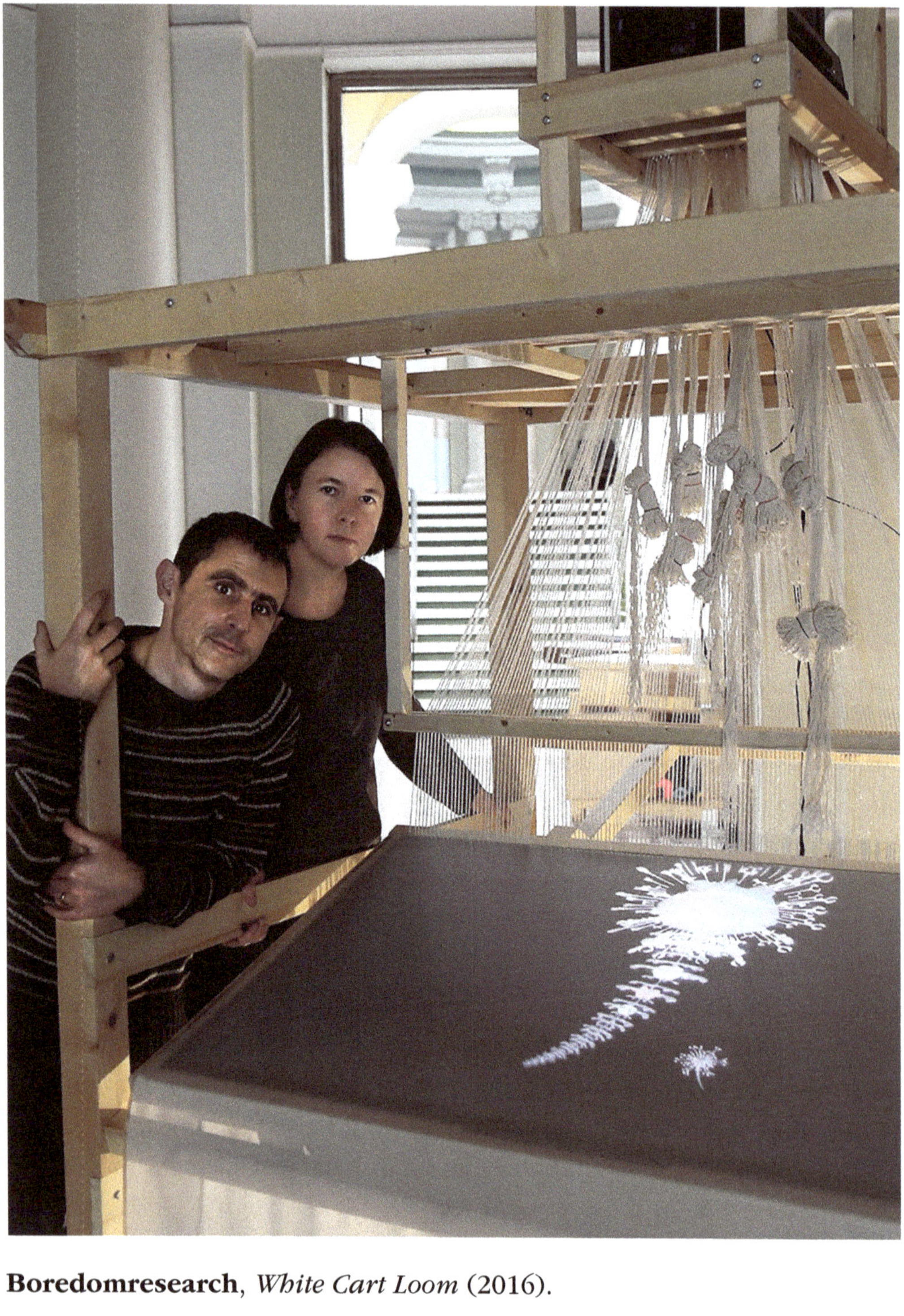

Boredomresearch, *White Cart Loom* (2016).

Imagining New Life Systems: Consistency Touched by Chaos Boredomresearch

To successfully simulate life is thought to hold untold promise for making a better future. A creeping unease, however, emerges that a simulation can only ever be incomplete. In this interview with the artist duo boredomresearch, we explore how our increasing confidence in modelling living systems is only matched by our inability to fully understand the consequences of such actions.

A central, recurring theme in your work is the matter of life's complexity, understood through its astounding diversity and shifting – but ongoing – coherence: life at the transition between predictable order and randomness. What is the appeal of such complexity?

The excitement of 'consistency touched by chaos' can be seen in the coherence which binds diversity. In psychology, the consistency principle describes a strong psychological need to remain consistent with prior acts and statements. This might also describe the point where new ideas create a dynamic tension between a need for familiarity and the possibility of change. In contrast to a myth, propagated by many historical narratives describing revolutions in art and science, the need for familiarity and the possibility of change both appear subject to a principle of consistency. Here, abrupt changes are resisted, even resented. It seems apt then that reward should be found in that which challenges our expectations without destroying them. Such a balance is mirrored in natural diversity with many small differences providing a wealth of variation. For us, complexity is the pleasurable overlap of these opposing forces, where complicated interconnecting parts provide an intoxicating sensation of the familiar spiked with the extraordinary.

Our artwork 'White Cart Loom' (launched in November 2016) captures this perspective. It takes the form of an early 19th-century weaving loom, drawing inspiration from the Jacquard loom of 1804 (the first programmable machine). A length of fabric, as though in production, provides a surface for animated pearlescent forms to materialise from where a digital shuttle shoots back and forth. These life-like forms are inspired by an ancient teardrop motif of Persian origin, known in many parts of the world as the Paisley pattern after the name of the town in Scotland where textile production took place that incorporated the motif in their designs. The animated forms swimming across the fabric surface are inspired by the freshwater pearl mussel, *Margaritifera margaritifera*, now a critically endangered organism that was once prolific in the White Cart River which winds its way between Paisley's former textile factories. The artwork weaves a narrative combining current scientific and ecological data in a fight to save this rare organism, now locally extinct. To these ends, the 'White Cart Loom' uses computation to enable the creation of 7.3 billion unique life- like forms, one for each human alive on earth at the time of launch. Considering the affordances of contemporary technology, we essentially ask through this project: 'How should we value the unique and last representative of a living species'?

Your approach is not based on making interventions into life *per se* but, rather, to simulate it in a way that brings different forces and pressures to the surface, so opening them up to inquiry. What are some of the experimental and investigative techniques you have come to use?

Research funding ensures the enduring importance of an intervention-based approach to life, and it remains central to the scientific endeavour.

Researchers are expected to deliver world-changing powers for our benefit. All life is subject to this ingenuity. Recently, we were based in an artificial life lab at the Karl Franzens University of Graz, Austria, where we collaborated with scientists aiming to install the world's largest robot swarm in the highly polluted environment of the Venice Lagoon. With an interest in creating bio-inspired control systems for robots, the team consists predominantly of biologists who see swarm intelligence as a robust engineering paradigm. In science, simulation often helps illuminate a specific problem, and in the a-life lab in Graz, the honey bee has provided a good research model for understanding the value of distributed intelligence. We were shown an example of simulated agents equipped with artificial hormones; here, the simulation provided a basis to evaluate the enhanced seek-and-consume abilities of bio-inspired control systems.

Simulation is also significant for creative arts practitioners. For example, in the animation and special effects field, visual qualities like fluids, cloth, or even crowd behaviour are often synthesised in this way. We too use the term, at least casually, in relation to our own creative practice, though we also like to consider simulation as an expressive process. In science, simulation helps test ideas through the careful application of a focus that excludes unnecessary or irrelevant detail. This process of reduction is similar to that exploited by the artist, by which a particular idea or interest becomes central to a study. In contrast to science, however, the models we find interesting have expressive potential. Simulation extends notions of the mechanical to the aesthetic. Here, uniquely afforded creative gestures transcend an inadequate representation of reality to inform our understanding *and* experience of life: Considering the lab's simulation of ravenous, hormonally enabled robots, we are reminded of human patterns of consumption, patterns that are now widely understood as the predominant force shaping life on earth. In this context, our chosen role in the lab – ignoring robust engineering metaphors – was to consider both the fragility of the swarm and the importance of hormonal influence on negative emotions amongst swam members. In our opinion, these negative forces should not be ignored in an attempt to understand swarm behaviour because they may play a crucial influence in how it copes with future scenarios less favourable than the current. More so, as our ability to intervene in system behaviours increases, our ability to evaluate the consequences of our own actions is exceeded.

A second core theme in your work concerns how the act of simulation – in rendering open to intervention the mechanism and substance of living systems – challenges the boundaries between disciplinary practices. Where has your work begun to open up such disciplinary conversations?

Photography introduced the icy indifference of a camera lens, indexically linking points in physical reality and the imagined universe of the image. Before that, the capture of the elusive essence of life (spiritual, biological, and physical) in visual form was the unique domain of the artist. Despite a highly developed appreciation of the artistic affordances of photography, concerns over the absence of human spirit in contemporary digital arts practice remain a prominent point of discussion. In the domain of computer graphics, a virtual camera captures a virtual world, which is rendered in visual form as the result of a simulation of light particles bouncing from surface to surface. This level of abstraction is more established in scientific fields. Here, models that are based on data collected from the real-world experiments, and subject to a form of disembodiment, furnish society with knowledge concerning the effects of possible, real-world interventions. Therefore, rigour in the scientific process requires that the integrity of a model is constantly challenged in relation to the measurable physical and biological universe it represents – in any case, it should not be influenced by the subjectivity of the author.

In our project titled 'AfterGlow' (2016), we collaborated with a mathematical modeller working in the field of epidemiology to create an artistic expression of an infection transmission scenario. Although the visual expression that resulted was very different from that used to communicate scientific insight, the underlying model was similar to that which might form the basis of a scientific inquiry. It subsequently became clear to us that there was more in common between artistic and scientific practices than might at first have been apparent. Both artist and scientist employ technology to create powerful abstractions which intensify a particular area of interest. A valuable rendering is then created to share the significance of the underlying process with individuals who bring to bear their own experience and interpretation. In artistic fields, there is a greater acceptance of differences between interpretations, while, in science, a singularity of meaning is enforced by strict protocols which aim to ensure immutable translation. Individuals lacking the necessary key to unlock this value remain outside its field of influence. In our opinion, much science communication fails to recognise the value of art in providing polysemous expressions with which the growing disconnection between expert and lay person can be overcome.

In `White Cart Loom', you explore the variability of shell formation in the freshwater pearl mussel as refracted through different biological, social, cultural, computational, and economic lenses. In what way do you see this project – and others from your work – as reconfiguring the relationships we traditionally see between these different forms of activity?

The value of the freshwater pearl mussel has been recognised for centuries, primarily for the beauty of its unique pearls. Despite the ease by which pearls can be farmed and synthesised artificially, there remains a demand for them, encouraging illegal poaching of this critically endangered species. Filtering around 50 litres of water a day, scientists highlight the importance of mussels for maintaining water quality over their commercial exploitation. The teardrop shaped motif, central to the textile industry in the town of Paisley, gives visual form to a reverence for nature but one lost in translation. Imported from the Middle East, the pattern's exploitation contributed to the local extinction of the previously abundant freshwater pearl mussel. The riches the design brought to Paisley are expressed in extravagant, now crumbling, architecture – a wealth long-since spent. The loss of natural diversity remains an enduring cost.

In our project 'White Cart Loom', we revisited the concept of the programmable loom (the cutting edge technology of Paisley's industrious past), so celebrating this first recognition of the creative significance of programmable technology. Although the loom's contemporaries favoured its capacity for wealth generation, the programmable loom allowed for the exploration of aesthetic variation through re-running programmed patterns with different colour schemes. It is this less-considered affordance that suggests, to us, a more important focus for human cultural innovation recognising the importance of diversity. Increasingly, computational technologies provide the tools to negotiate the complexity of ecological systems. This means we are now well placed to move beyond a reductive approach to, for example, food production or environmental management that favour standardised units of production and intervention. Where food crop monocultures have been maintained through chemical warfare, these can now be replaced by complex tapestries of interacting parts. Where environmental simplification and reduction has been valued because of its short-term benefits, such value is to be outweighed by the riches of investment in longer-term biological diversity – diversity that is itself reflected in the richness of global cultural diversity.

The way in which your practice folds together and reconfigures different influences in the study of life systems opens up new imagined (but previously inaccessible) possibilities. What are some of the decisions that lie behind this process of reconfiguration, and where have surprising outcomes emerged?

Surprise and process are the primary reason we choose programming as the medium of our work. Many think of computers as machines that follow instructions, precisely and without error. While true, the nature of those instructions may incorporate complexity in such a way that our expectations are also challenged. Our preference is for something that is evocative of the richness we perceive in life. Vast creative spaces are revealed at reasonably low thresholds of complexity. The overwhelming diversity present in nature represents only a tiny slice from the space of possibilities. What of all the life living but undiscovered, lost but unrecorded, or even that yet to become? Many of the artworks we create share this quality in that most of their content will never actually be seen: Their rules allow for an enormous – often un-witnessed – diversity to be generated. Many creative decisions are made in steering this process through which we aim for maximum freedom in what is created, while maintaining artistic and functional integrity. Our reward is to experience the surprise of unexpected, emergent forms. The most surprising element, however, is how difficult it is to synthesise this freedom without catastrophic collapse. To us, this stands as a blunt reminder of the wider limitations in any attempt to manage the complexity of biological systems.

From one angle, a simulation of life processes is self-contained (indeed algorithmically deterministic) in a way that life fundamentally is not. From another angle, however, your works deeply embed such simulations into contemporary `living' contexts, behaviour, and activities (such as in galleries and museums); is this where life lies in your work?

Ignoring the celestial energy from the sun, life on earth is predominantly self-contained. The sum total of all the earth's constituent ecosystems is immeasurably more complex than any simulation of it. In creating a simulation or model, we may imitate an existing system or mechanic, but we also create something new – a new expression that is subject to its own rules and limitations. In effect, we create a new universe connected to an outer, sun-like, energy source. Many of the works we create use algorithmic processes that, although deterministic, produce operations that are impossible to fully predict until they are computed. This, in our opinion, breathes life into the work allowing the viewer to enjoy the sensation of surprise as changes occur. Life is also an ongoing process of change to us; computation is the best medium to express this. In a scientific context, an urgent need for results applies a pressure to the modelling of systems, encouraging the use of computation to accelerate simulated time, enabling, for example, future system outcomes to be predicted in advance – often with a view to making a positive intervention. Currently, the use of abstract computational representations of natural systems is more common in science than art. Our concern is that this creates an uncomfortable power relationship, whereby the non-expert citizen has little basis from which to believe the insights of science, other than to accept their own ignorance. We would like to see algorithmic expressions of life become more common in a wider cultural context, such as in galleries and museums, to address this. Through our contribution to this debate (in the form of works such as 'White Cart Loom'), we hope to help form a common aesthetic understanding of these simulative processes, aiding a positive synergism whereby art, science, and society can move beyond a current state of discord in relation to our sustaining environment. Earth is, after all, a self-contained process that can only be run once.

Taking this further, how would you wish audiences to place themselves in relation to your works? Are they to be part of a didactic process, or is there a route by which they can feel themselves into the life of your works, a way of `becoming media'?

Scientific datasets can strike the uninitiated as being destitute of vision. As artists, we seek poignancy, not to overwhelm an audience with facts but to make visible an undercurrent of essence that has significant societal implications. The vehicle of art can, at least, broaden an audience's reception of such data, formerly numbed by its deliberate, anti-emotive language, so provoking intrigue and emotional connection. For us, the distinct didactics of art present powerful tools to synthesise responses that are different from insights. In addition to a conceptual formation of meaning, favouring an immutable symbolic communication that fails to completely capture the extraordinary nature of life as experienced by the living, an artistic expression provides a missing visceral dimension. Science is currently the predominant paradigm through which technological creation brings into being the mechanical basis of our daily lives. Increasingly, the mechanical basis of life is understood and manipulated with a technological mindset formed in the industrial revolution, favouring standardised units of production like that seen in palm oil plantations. This and similar agricultural innovation continually erodes the habitat of our closest biological cousins (Borneo's Orangutan population, for example, has dropped by 150,000 in just 16 years). As a consequence, we have become a living expression of a disconnection between what we know and what we feel. Experiencing a world increasingly limited by the outdated ideals of mass production, discomfort is felt by many when the benign tasks of buying food forces them to either ignore, deny, or negotiate food chains that reek environmental destruction at a distance. For human culture to regain its integrity, we must both feel and understand the material basis of our world. In effect, how we want to feel about the world should inform the technological basis for our lives, not the other way around. Our artworks are a response to the mechanics of natural systems, their scientific understanding, and the wider concerns we face at the level of the everyday citizen trying to get the best from life.

Imagining new life systems requires an appreciation of different ways of knowing and conceiving; it also requires an understanding of where weaknesses lie in one's own grasp of the models through which we know and conceive reality. How did you get here, to boredomresearch?

A contradiction inherent in 'insight' is that it consumes the necessary ignorance from which we can conceive anew. To understand darkness, we must first turn off the light that obscures it, freeing ourselves from the constraints and limitations of what we feel we know. Although we can only think in the shadow of our mind, the rigour of research practice demands we expose the mechanics of thought through methodologies that bleach the bright colours of playful freedom. But play may be more important than we care to acknowledge. In engineering, there exists a theoretical ideal to remove all wasted motion from moving parts. This 'unnecessary' motion, often referred to as play, is in reality essential for movement. Without the freedom of play, the machine literally seizes. Although the mechanisms of research may be expected to steadily fill gaps in knowledge, replacing doubt with certainty, for us, is to become stuck in an unchanging world – to become bored. Disengaged with one's current environment, while maintaining an uncomfortable fidgety energy keen to act, boredomresearch aims to escape the limiting friction inflicted by the certainty of established academic structures. Boredom is a force providing insights liberated by imaginative freedom where the illusion of rigour gives way to what may, or may not, be possible. As new and imagined life systems become reality, we should remain mindful of the impossible ideals of systems without play and to the impossibility of exactitude. To achieve this, we must temper the actual with the imaginable; only then can we be sure to provide a better situation than the current.

As humanity invests a significant proportion of its creativity in the endeavour of resolving problems arising from rapid population growth, we bring to bear the sum total of our knowledge. This base of understanding has predominantly been built in the image of deeply ingrained cultural beliefs, limited by unquestioned assumptions – healthy not sick, rich not poor, easy not hard, more not less. In the early 19th century, American philosopher Henry David Thoreau, famous for his reflections on simple living in nature, rejected the predominant cultural wisdom of developed society. Foreseeing that one has established the basis of their life 'When one has reduced a fact of the imagination to be a fact to his understanding ...', he explored the possibilities of life outside the constraints of developing technological innovation; these he perceived as encumbering the freedom of humankind. He pitched himself against nature's adversity to find life's true and essential needs. Unsatisfied with developing market forces and mechanisation, he recognised that 'man's labour would be depreciated ', leaving him 'no time to be anything but a machine '. In response, he sought a visceral experience of both the nourishing and antagonising forces of nature, from which he foresaw the methods and insights of ecology. In the present, as we make use of a recently gained mastery of living media, underpinned by all that we know we know, we should also consider the Confucian 'unknown unknowns' that can only be sensed by a free imagination. Subject to the darkness of our knowledge and the light of our creative freedom we should, at least, observe Thoreau's observation that 'The finest qualities of our nature, like the bloom on fruits, can be preserved only by the most delicate of handling '. In imagining new life systems, we must recognise that life is, and should remain, fragile.

Author Biographies

Boredomresearch is a collaboration between British artists Vicky Isley and Paul Smith. Through computational technologies and real-time animated environments, they ask questions of complex system behaviours, exploring their robustness, sensitivities, and vulnerabilities. Often working in collaboration with scientists, their work has explored the biological signatures of neural activity, the frontiers of disease modelling, and our cultural obsession with speed. Boredomresearch has worked in collections around the world including the British Council Collection and the Borusan Contemporary Art Collection. Recent international exhibitions include Artience, Daejeon (KR); ISEA, Manizales (CO); Data Aesthetics Exhibition, Amsterdam; Bio-Art, Seoul; and TRANSITIO MX_06 Electronic Arts & Video Festival, Mexico City. More information on Boredomresearch can be found at www.boredomresearch.net

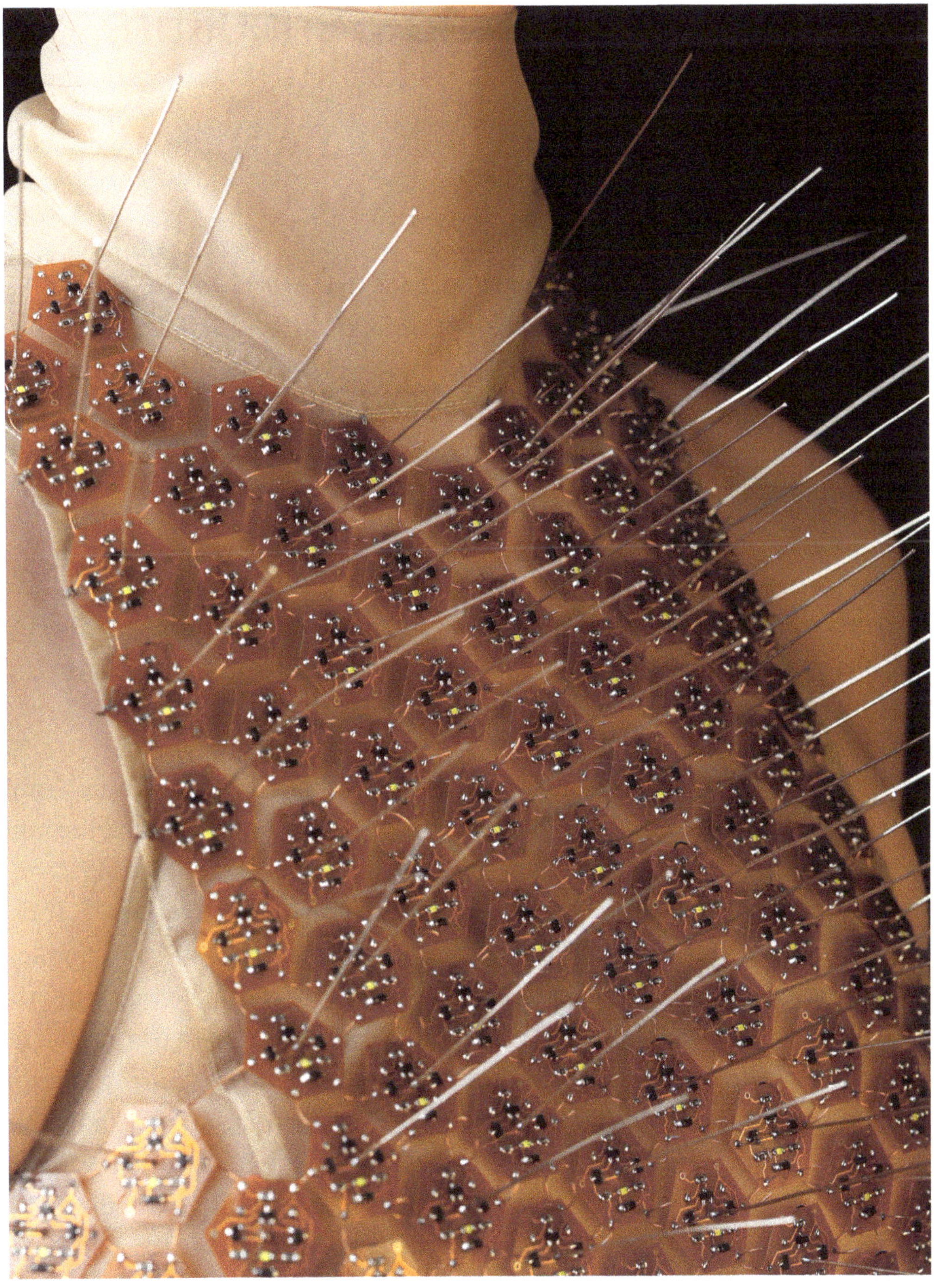

Vivian Xu, The Sonic Skin (2018).

Radically Rethinking Sericulture

Vivian Xu

The individual is always in interaction with her environment, a coupling that enables different identities to stabilise over time – the silkworm and the mulberry tree. Non-life also features in these couplings, but its status in the relationship is far from clear. Where does non-life end and life begin? In this interview, Vivian Xu asks what it means for technological systems to be understood as a natural part of this mix.

We are more comfortable thinking of living systems and mechanical systems as separate – life versus non-life. In the realm of interactions between them, where might a different understanding of that relationship be possible?

A large part of my practice is based on the study of the machine –animal continuum. Humans have long explored conceptions of life through technology (for example, through mechanical automata imitating living systems), but what intrigues and inspires me in a contemporary sense is a relationship between mechanical and behavioural systems (chemical and biological) that interface the unpredictability of the living with the controlled (predictable) behaviour of the machine. It is one that complicates our understanding of both – of the machine, and of life. In my eyes, it is electricity that unites them in action and interaction: Digital machines operate through electronic circuits, electricity serves as a medium for digital communication, and, in the case of the biological body (whether it is the nervous system or DNA), electricity acts as a medium for biological communication. It is electricity – as medium and communication – that breathes life into both.

So, I find myself adhering to a more bionic view (electromechanical, if you will) of the relationship between life and non-life. I am especially inspired by Manuel DeLanda's vision of 'Nonorganic Life' (1) that tries to re-examine that relationship. What I draw from DeLanda's work is that the boundaries of life, or categori sation of life, is subject to fluctuation and open to challenge: This is especially true in today's world. In this understanding, our concept of life changes in reference to the perspective of the subject, i.e., what counts as life is intimately bound to the eye of the beholder. What we perceive as life often exists (and transforms) in a similar timescale to us, and so, because of this, appears life-like. This is the reason why we perceive plant life to be 'more alien' than mammalian life. If, however, we could set up a camera to capture the formation of geological landscapes over time and play that captured footage back at high speed, we would discover that they too exhibit behaviours that are life-like. They appear to self-organise, can be highly active and generative, and are open to change from interactions with other entities. Matter is constantly changing and rearranging in time ; so it is just a matter of whether we are able to perceive it or not. Eastern thought emphasises concepts that are similar to this – ideas which are becoming more and more prevalent in contemporary western philosophy.

How do these ideas around the machine – animal continuum find expression in your work?

My work 'Living Devices' is the first of several explorations dealing with hybrid systems, where the system relies on both parts as a whole to function and generate meaning. Here, the device uses the electricity generated by embedded circuits to control a petri dish environment, generating a changing electromagnetic field that modifies the growth of bacteria into different patterns. The circuits are simple, running electricity between two node clusters to form a closed circuit through the agar (using the agar essentially like a wire). But because the agar is a conductive medium, the path of the electrical current is more unpredictable than a wire, creating an electromagnetic field around the nodes. Affected by the field, bacteria may grow or not grow in different sections of the petri dish. Paired with various seeding designs and patterns, different end results can be generated. Originally, I worked with *E. coli* for practical reasons, but it would be interesting to continue this project with more exotic bacteria that have a greater sensitivity to electricity.

'The Silkworm Project' develops these ideas surrounding machine design and life further by working with animals that, themselves, have a more productive and interactive relationship with their environment. As part of a larger body of work called 'Insect Trilogy', I have been looking at three insect architects (silkworms, ants, and bees) with the aim of designing machine environments and a machine logic that create an intelligence system different from our (and their) own – a new bio-electronic ecosystem. 'The Silkworm Project' poses questions around production and autonomy in the designing of machine systems, creating a machine environment in which the spatial perception of the silkworm is hacked, causing it to spin self-driven, organic, three-dimensional silk structures.

Activity in these `new ecologies' emerges in time, meaning that its different components – the silkworm and the machine – must interlock purposefully at each step. How does time play a role in this work?

As a former film student, I am particularly interested in the nature of time-based media. For me, a biological medium is a time-based medium; but whereas film unfolds temporally within a VR, organic life unfolds over time in the physical world. A recreation of organic life is a recreation of an 'all together' time –space reality within an organism. Much like the internal film time inherent in the virtual world of the screen-based narrative, bio medium also has its inherent time – the circadian clock or biological clock.

Inherent time in the organism provides the scale on which the experiential reality of the organism is generated. When growing silkworms, time plays a crucial role in the determination of the worm's life cycle. Worms hatched in the beginning of April grow larger and have longer cycles than worms hatched in the beginning of May. This method is, perhaps, a rather simple and crude way of manipulating time in life-forms, but, if we consider the rapid developmental speed of current biotechnologies, there may very well be technologies in the near future that can change the 'frame rate' of living beings. Though it is near impossible to understand the internal reality of a worm (or any animal), one can speculate on how such technologies might change the experience of time for an organism and introduce an alternative sense of reality.

What, then, is an artist language based on the manipulation of biological media? What, then, is meaning in bio media? My goal in 'The Silkworm Project' is to try and negotiate between the biological time of the organism and the technological time of machines in order to find an equilibrium between the two. New realities revealed through new tools bring about new challenges of perception. Accordingly, we need to adjust our understanding of the world to better reflect the tools (both physical and conceptual) we use to generate new understanding. The purpose of redefining our definition of life is, therefore, to reflect the new realities that have been exposed. To hold on to historic models of perception is like trying to solve modern-day crises with Renaissance toolkits. Or, worse yet, blinkered by old models of perception, we may fail to foresee new challenges that loom immediately ahead. With 'The Silkworm Project', yes, I am interested in how technological and biological systems can generate a new coherent 'whole', but I am also interested in how we might play with the historical logic behind the development of computational and digital technologies. While the start of the information technology age was strongly influenced by the culture of weaving and textile production, I want to use digital processing in my work to influence the organisation of silk production straight from the silkworm's own mouth. This is a different way of modelling the relationship between technology and life through new types of machine life that lock them intimately and coherently into each other.

In `The Silkworm Project', an individual silkworm is faced with a new environment – one shaped by new technological parameters. You have suggested that a unique internal logic emerges in this new ecology. Is that logic one of an experimental `disorientation' or `adaptation' for the silkworm?

This is not so easy to answer. Originally, I worked on creating an electro-stimulation grid that could ideally both stimulate specific animal behaviours and support a successful spinning environment. Here, the silkworm would be acting as both the input and output of the system. The end result, however, was a machine that was not able to properly function. For one, although silkworms are able to respond to electro-stimulation (because it can tap into their nervous system), I was unable to identify an ideal current range where a desirable behavioural reaction could be triggered in the worms without harming them. This forced me to look at spinning behaviours in a new way. I adjusted my approach from designing with the silkworm (i.e., using the silkworm as a tool influenced from outside within its environment) to thinking about how to design *for* the silkworm.

I started conducting my own spatial spinning experiments with silkworms, looking at how the insects navigate through space individually and collectively. For the collective experiments, I cultivated silkworms that produced multi coloured silk using both the Singaporean method, based on feeding coloured feed to the worms, and the Japanese approach of genetically engineering silkworms. Through colour tracking methods, I was able to observe the negotiations of two worms spinning in a common space and building upon each other. It was surprising to find that there were very few errors or overlap in their collaborative silk spinning, with the spatial territory of each worm clearly marked through colour differentiation. Using basic properties of their own bodies (morphology, size, and shape) almost like a measuring stick to help predict and understand their environment, paired with a building process honed over thousands of years of evolution, the silkworms bypass a top-down spatial blueprint methodology in favour of a bottom-up responsiveness to local conditions – one that can generate an ever-changing array of spatially expressed silk-spun forms.

The goal I then set myself was to disrupt this equilibrium of insect perception, i.e., to introduce a new environment that generates spatial blind spots – a property that can be harnessed to create new types of spun-silk structures. This new environment includes a glass chamber where the curvature surface of the glass prevents the fully developed healthy pupa from identifying the corners and angles it would normally use to build a three-dimensional framework for its silk construction. The size of the jar is determined by the size of a healthy pupa, where too-wide a circumference would result in flat silk weaves, and too-steep a curvature would result in a fully formed cocoon. A vertical spinning motion of the chamber affects the silkworm's sense of gravitational pull, where the slow spinning provides a constant change of gravitational direction, thus confusing the insect's spatial orientation. It is essentially a machine that reflects the space beyond the silkworm's own perception. Though my experiments may, at times, yield interesting results, they are often hard to replicate.

To understand sericulture, we need to look beyond the immediate ecology of the individual silkworm. Can you tell us a little more about the wider sericulture ecosystem and how it brings together living bodies, technology, and human culture?

The relationship between Chinese people and the silkworm is very complicated. In our history, the advent of sericulture came before the invention of the written language. Its beginning is often attributed to the first Empress of

China: The story goes that she was sitting under a tree drinking steaming tea one day and a silk cocoon fell into her cup and unravelled into a continuous strand of silk. Though a romanticised myth of the beginning of silk reeling, rather than historical fact, it does go to show just how ancient this industry truly is. Up until the European medieval period, when sericulture started to spread across Europe, silk production was still largely centred within China (with the exception of Japan and India). As a luxury product, it was the basis for economic exchange between East and West via the Silk Road. Early weaving technologies (like the Jacquard loom) went on to inspire the invention of the computer.

Those new to sericulture may fail to realise the immense human labour needed to care for hundreds or thousands of silkworms. In the summer of 2019, I raised a total of 600 worms in Berlin divided into three batches of different age groups. I raised these worms from eggs (the size of a sesame seed) until they were fully grown worms (roughly 7 –8 cm when healthy and well-fed). I spent roughly 4 –5 hours daily feeding, cleaning, and documenting the worms. I needed to plan my daily routine – meetings, outings, etc . – based on the silkworms' feeding and cleaning needs. In this instance, the silkworms are more in control of my daily activities and timeframe than I am of theirs. In my first-hand experience, I would say this is first an industry of human and technological labour, one built to serve the needs and capabilities of an insect species, where the timeline of the insect dictates how that labour is organised: It is more of a socio-ecological system than many would imagine.

My focal point for 'The Silkworm Project' begins with the historical intersection between the organisation of material culture and the organisation of information and data. I see myself as following the traditions of both. Rather than working to change or replace an age-old tradition, I want to understand what the drivers are behind this extraordinary relationship we have created between living organisms, technology, and human culture. What interventions are possible as a stimulus to re-thinking those relationships in new ways? In both China and the West, agricultural treatises played an important role in introducing sericulture to a wider audience. I am currently working on an artist book that models itself after these manuscripts in exploring the complex social, biological, ethical, and political issues that have come up in my research of this project.

Sericulture is able both to resist change and adapt to new ideas. As the silk industry continues to innovate, how is the ethics of sericulture changing, and what does that mean for this ecosystem perspective?

Sericulture is built on the killing of silkworms, but, because the practice is also almost synonymous with Chinese culture (and as old as Chinese culture), there is an important layer of heritage and emotional attachment to the silkworm that is deeply rooted in Chinese society, even today. The ethical debate of sericulture stems mainly from a Western point of view of humanitarian practices. There is humane silk farming, which uses the Indian silkworm, but the silk produced is more like cotton and, therefore, is not as fine as silk produced by the Chinese silkworm (Bombyx mori), which still accounts for all the luxury silk products we consume. Although it is easy to say that old production techniques should cease and new ones that adhere to a Western sense of ethics should be embraced, critics of the method fail to understand the meaning of silk making in China as social and material culture. The whole-sale adoption of new methods risks being both reductive of that culture and impractical to implement. It would be a change that would affect the industry across China without taking into account the perspective of generations of Chinese family businesses that have thrived using older techniques.

In a way, this reveals a critical difference between how Eastern and Western thoughts relate to concepts of death. While the focal point of Western ethical debates on sericulture circle around the binary of life versus death,

I think traditional sericulture industries in East Asia have the quality and purpose of life as its main concern. Ancient sericulture treatises in China and Japan lay rules for taking care of silkworms as well as rules for the behavioural conduct of silkworm farmers. The worms are boiled in their cocoons towards the end of their life cycle; in China, the pupas are not wasted but rather cooked as high protein food, even today. When these practices are seen only in fragments from an outsider's point of view, it is easy to label them as superstitious, cruel, or abnormal, but it makes more sense when you view the ecosystem as a whole. Compared to other agricultural models used today, I find this more of a sustainable approach. As described earlier, 'The Silkworm Project' tries to embrace a different logic within this debate by exploring how new, meaningful relationships are possible that take into account the biological, cultural, and social elements of the wider sericulture ecosystem.

A recent extension of your project addresses how new `live' printing technologies might revolutionise the production of garments and the field of wearable technologies in general. What insights are you beginning to uncover, and can you speculate on new forms of cultural- and self-expression that may emerge in the future as a consequence?

'The Silkworm Project' got me thinking about wearables and how they might help us redefine the realm of our bodily relationship to clothing, even perhaps to reframe our bodies entirely. Since last year, I have been working on a wearable technology series that looks at skin as an interface, speculating on how we might use it to explore new sensory ecologies. Skin is particularly fascinating as the boundary between our internal and external environments – between ourselves and others. The idea is that by expanding, even blurring, your senses at your 'natural' boundary, you can momentarily increase your perception of the world around you. In a similar vein, my collaborator at Dogma Lab – artist and musician Benjamin Bacon – has been working on body implants and instrumentation to exploit our capacity for enhanced sensory abilities. He has, for example, embedded magnets into his fingers that allow him to sense electromagnetic fields around him, such as that around electrical wiring in a wall or around an electrical socket. My interests are more in non-invasive modification and how we might learn from – and mimic – the sensory systems of other animals that are alien to the human experience at present.

'Electric Skin' is the first of two wearable pieces I am developing and draws inspiration from many animals' ability to sense electro-magnetic fields in their environment. The approach I am taking is to map this sensory function onto a circuit armour hosting antennas, which would allow human wearers to use their skin to experience their technical space, i.e., the layer of reality they cannot normally perceive around them made up of electrical signals. Our immediate environment has changed dramatically over the past 100 years with the development and proliferation of radio and information technologies; I think there is no reason why we should not try to 'evolve' and keep up with those changes. Recently, I was able to test a small patch of the circuit 'fabric' that I had created with users. It translates minute electromagnetic signals from the environment to your skin via gentle vibrations from vibration motors. The second project – 'Sonic Skin' – takes the idea of elevating human sensibilities with the assistance of wearable technology in a different direction. Inspired by a bat's or whale's sonar system (where the journey and reception of sound bouncing off surrounding surfaces is used to illustrate the spatial relationship between animal and environment), this project will develop a wearable armour of audible and directional sound that projects into (and back from) the environment around the contours of the wearer's body.

You are the cofounder of Dogma Lab, along with Benjamin Bacon. The lab is set up to enable the different creative and research activities your work depends on. What does this space mean to you, and what does it say about the need to work in new ways around complex topics such as human sensory futures?

Dogma Lab is really a personal playground for Benjamin and me. It has two components – a commercial side and the non-profit experimental side. Both Benjamin and I have worked in different realms such as experimental art, music, community, academia, tech, and commercial design, and we find it extremely important to be able to bring different perspectives into new projects. It gives us the opportunity to learn-through-doing and, in turn, to offer that experience of enrichment to others. In this sense, being elastic and multifaceted in the way we work is what we really love about design as a field of practice. Looking to the future of Dogma Lab, we are trying to build up a network of trusted professionals and collaborators that draws on previous project partnerships. With this, we are trying to create a healthy creative ecology that allows the resources obtained from commissioned projects to fund further experimental and research-based work. It also means we can use experimental work to inspire new ideas in the public realm through the creation of more interesting products, experiences, and communities. Right now, we are just beginning our journey in that direction. Hopefully, we will be successful.

I am a strong believer in education. In today's world, learning is shifting away from universities towards a more decentralised system, where one can gain experience and professionalise via multiple platforms, institutions, and organisations that exist independent of traditional schooling systems. I think that my experience at Genspace was extremely important in this sense, in that it opened my eyes to new possibilities for educational and collaborative practice. At the same time, having spoken with innovators and community organi sers in parts of Europe and East and Southeast Asia, one finds that each space is run in very different ways. Depending on the local social and cultural atmosphere, each strives for its own – often very different – goals. There is little value in making general statements across these spaces, as each is tuned to the needs and interests of their own communities.

For these reasons, we are supporters of the DIY community; it is a means of democratising access to knowledge, tools, and skills (in a way that has not been possible in the past), whilst also responding to local conditions. The pool of creative talent encouraged to engage with any number of subjects through these communities is amazing. But this is not enough. DIY practitioners will need to gravitate towards more critical and systematic methodologies of creation if they are to gain a deeper understanding of a subject, so progress from enthusiast to expert. We are very fortunate that design methodology in the 21st century allows for both of these – a means of treating complex issues through critical design approaches while embracing an openness that allows for experimental collaboration (and the absorption of other perspectives). Design provides a basis for us to approach the world, but, following the same argument, it does not offer 'the solution' to everything. It must respect and engage with other disciplines to truly create impactful results. Therefore, we advocate for a trans disciplinary approach to collaborative creation over an interdisciplinary approach, i.e., one where new knowledge systems and processes are generated over long-term partnerships rather than just drawing on different bodies of knowledge to create something new.

Reference

DeLanda, Manuel. Nonorganic Life. In *Incorporations*, edited by J. Crary, and K. Sanford, 128 – 167. New York: Zone, 1992.

Author Biography

Vivian Xu is a Beijing-born media artist, designer, researcher, and educator. Her work explores the boundaries between bio and electronic media in creating new forms of machine logic, speculative life, and sensory systems. Her work often takes the form of objects, installations and/or wearables. She has shown, lectured, and performed at various institutions in China, the US, Europe, and Australia; these have included the National Art Museum of China in Beijing, the China Academy of Art in Hangzhou, the Chronus Art Center in Shanghai, the New York Science Museum, Art Laboratory Berlin, SymbioticA at the University of Western Australia, and the Kapelica Gallery in Ljubljana. Vivian co-founded Dogma Lab, a trans disciplinary design lab in Shanghai that is dedicated to creating experimental research at the intersection of design, technology, art, and science. Her work has been featured in media and press, including VICE China, Elle US, and the China Global TV English Channel. Vivian received her MFA in Design and Technology from Parsons the New School for Design in 2013, and she is currently an Assistant Professor of Media Arts at Duke Kunshan University in China.

Index

Index of names

Index of terms

About the Editors

Arthur Clay has designed and implemented trans disciplinary events focusing on creatively connecting art, science, and technology within diverse cultural contexts in many parts of the world. As co-founder and artistic director of the Digital Art Weeks, his activities include developing and platforming projects that pioneer new technologies including working with renowned institutes such as EPFL, ETH Zurich, University of the Arts, Zurich, Sogang University, Seoul National University, and for various private and government agencies and institutions including SAST, Create Center at NUS, A2Star, and more. He has been supported by governmental agencies, art councils, private foundations, and industry partners such as Presence Switzerland, Pro Helvetia, Swissnex China, Swissnex Singapore, Japan Foundation, the Canada Council, LG Electronics, etc. He is renowned as a versatile artist working in diverse genres and has been awarded prizes for performance art, media art, music theatre, and composition. As an educator, he has taught at a number of high- ranking institutes in diverse parts of the world. At present, he continues to perform at international venues, is Guest Professor at Sogang University of Seoul South Korea, and directs the Virtuale Switzerland, a festival for invisible arts that promotes art using augmented, virtual, and mixed reality.

Timothy J. Senior is a cross-disciplinary collaboration specialist. His work focuses on how traditional forms of disciplinary activity in the arts, sciences, and humanities might be opened up to new collaborative interactions – a process of creatively challenging disciplinary values and methodologies to drive innovative lines of shared inquiry. He is co-founder and director of supersum – the wicked problems agency ; supersum supports new and unconventional partnerships to tackle shared problems that resist siloed ways of working (supersum.works). Tim holds an undergraduate degree in Biological Natural Sciences from The University of Cambridge, and post-graduate degrees in Neuroscience from The University of Oxford (DPhil 2008). He has held a variety of research and teaching positions at universities in the UK (The University of the West of England and Arts and Humanities Research Council), Germany (Jacobs University and the Hanse Institute for Advanced Studies), and the US (Duke University). Tim has published more than twenty peer-reviewed articles, edited-books, working papers, and reports. These span subjects as varied as medieval European urbanism, social neuroscience, the design of dementia friendly communities, and knowledge exchange in the creative and cultural industries.

Acknowledgement

We would like to thank Prof. Dr. Sunghoon Kim (김성훈 교수님) for his support over the years and for creating the opportunity for the artists featured in this book to exhibit their work to new audiences during the Seoul BioArt Festival; Prof. Jusub Kim (김주섭 교수님) for the many discussions around art and technology, and for hosting the 'wet media conference' at Sogang University which forms the core of this book; Prof. Robert Mitchell for his innovative and thoughtful work on the nature of media; the many artists who have shown us both the wonders of the world we live in and the abundance of life that inhabits that world with us; and, last but not least, we would like to thank the following for their assistance in bringing to life the various art and science projects featured in the Seoul exhibitions and conferences between 2014 and 2019: Stellar Park (박세진), Kevin Ahn (안재윤), Debby Kim (김보라), Heehyun Choi (최희현), and Tom Chung (정승기).

For Product Safety Concerns and Information please contact our EU
representative GPSR@taylorandfrancis.com
Taylor & Francis Verlag GmbH, Kaufingerstraße 24, 80331 München, Germany

www.ingramcontent.com/pod-product-compliance
Ingram Content Group UK Ltd.
Pitfield, Milton Keynes, MK11 3LW, UK
UKHW052000041026
15590UKWH00008B/2063
* 9 7 8 8 7 7 0 2 2 5 9 5 3 *